On the Wing

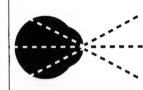

On the Wing

To the Edge of the Earth with the Peregrine Falcon

Alan Tennant

WHEELER PUBLISHING

Published in 2005 by arrangement with Alfred A. Knopf, Inc.

Wheeler Large Print Compass.

The text of this Large Print edition is unabridged.
Other aspects of the book may vary from the original edition.

Set in 16 pt. Plantin by Minnie B. Raven.

Printed in the United States on permanent paper.

Library of Congress Cataloging-in-Publication Data

Tennant, Alan, 1943–
    On the wing : to the edge of the earth with the peregrine
falcon / by Alan Tennant.
       p. cm.
    Originally published: New York : Knopf, 2004.
    ISBN 1-58724-898-0 (lg. print : hc : alk. paper)
    1. Peregrine falcon — Migration — Western Hemisphere.
I. Title.
QL696.F34T44 2005
    598.9´6—dc22   2004025224

For George

National Association for Visually Handicapped

serving the partially seeing

As the Founder/CEO of NAVH, the only national health agency solely devoted to those who, although not totally blind, have an eye disease which could lead to serious visual impairment, I am pleased to recognize Thorndike Press* as one of the leading publishers in the large print field.

Founded in 1954 in San Francisco to prepare large print textbooks for partially seeing children, NAVH became the pioneer and standard setting agency in the preparation of large type.

Today, those publishers who meet our standards carry the prestigious "Seal of Approval" indicating high quality large print. We are delighted that Thorndike Press is one of the publishers whose titles meet these standards. We are also pleased to recognize the significant contribution Thorndike Press is making in this important and growing field.

Lorraine H. Marchi, L.H.D.
Founder/CEO
NAVH

* Thorndike Press encompasses the following imprints: Thorndike, Wheeler, Walker and Large Print Press.

Contents

Part Three • The Bay of Mexico

Part One

PADRE ISLAND

I caught this morning morning's minion,
kingdom of daylight's dauphin, dap-
ple-dawn-drawn Falcon, in his riding

Of the rolling level underneath him steady
air, and striding

High there, how he rung upon the rein of
a wimpling wing

In his ecstasy!

Gerard Manley Hopkins,
"The Windhover"

1.

Partners

Trailing steady electronic beeps from the
tiny transmitter glued to the base of her tail
feather, the peregrine quartered tentatively
up Padre Island's barrier beach. A tundra
falcon, a barren-ground hunter born in the
Arctic, she had for the past two weeks
cruised with increasing randomness up and
down Texas's offshore islands, seemingly reluctant to leave the windswept flats for the
alien greenery of the mainland. But today
the springtime stream of tropical air welling
in from the Gulf of Mexico had gripped her
in its northerly flow, and after a final sideways cast along the surf she swept inland,
away from the sea.

"She's migrating," called Janis Chase,
the military attaché in charge of our radio-
tracking flight. "I think she's on her way!"

From 2,000 feet up, in the rear seat of
our single-engine Cessna Skyhawk I
watched the shoreline dunes give way to
bluestem pasture, and gradually the mag-

11

nitude of what we'd chanced upon sank in. With no companion, guided only by the ancestral memory she somehow carried within, our little hawk was staking her life. Nothing like the abstract idea of migration that I'd imagined, it was humbling even to be a spectator to the mortal intensity of what the tiny, determined speck below was doing. On that sunny coastal morning she was committing every ounce of will and spark of force that flowed inside her to a race toward home. I tried to imagine what could be going on behind her fierce eyes. Some inner vision, I guessed: a tucked-in cliffside ledge, set above the tundra, with long-unseen domestic details of surrounding rock and bluff-face. Maybe sounds, too: the whistle of arctic wind, and in the still air, familiar birdsong, the croak of ravens, or the scream of rough-legged hawks nesting nearby. No one would ever know what she thought, but it was clear that as we watched, something had swirled to life within this falcon, becoming the driving force of her entire being.

At the time, during the mid-eighties, it was thought that she would travel northwest from Texas, seeking the high-altitude passes of the Rockies, then work her way up the spine of the Continental Divide.

But only she knew the route she would really take, or where it might eventually lead. And only she would ever know if that knotted force of will — the power that was now sweeping her along at mile-a-minute speeds — would be enough to carry her there. Enough to lift her up, buoy her over the third of the planet that lay to the north, and then deliver her, perhaps weeks from now, onto the coal-shale crags of the Arctic. There, more than 3,000 miles from this humid Texas plain, on some late spring day the ledge where she herself was born might once again appear beneath her wings.

"That's enough," Chase ordered our pilot, George Vose. "I've got my departure vector; it's all we need."

Chase bent to her U.S. Army Chemical Warfare clipboard, whose logsheet had a heading for "Migration Route." On it she recorded the date, the weather, and the north-by-northwest compass heading that this falcon, the last of seventeen transmitter-bearing peregrines she had been assigned to monitor, had chosen as it left the Gulf's barrier islands. While Janis wrote, for a few more minutes Vose flew on. I could see that — even though we knew her mostly from her relayed transmitter signal

— it was hard for him to leave the falcon flickering along below, now embarked on a quest toward her unknown, almost inconceivably distant home.

Then Janis looked up and in irritation motioned for George to go ahead and turn. Reluctantly, he banked us slowly around, descending across the bay to Cameron County Airfield and what had suddenly become a far smaller world.

Smaller because, even with my daily chance to capture and band peregrines as part of a peregrine research team, seeing the metamorphosis that migration had brought about in one of the falcons whose kind I'd watched, and even captured on the tidal flats, was astonishing. Afterward, Padre's peregrines were no longer simply beautiful raptors to be lured in for banding. They were part of something larger. Something ancient and powerful. Global as the tides, though at the moment all I could think of for comparison was gazing, as a kid in the port of Houston, at ocean-going tankers from Liberia, freighters from Singapore and Seoul, Buenos Aires and Dakar — places I knew about but never hoped to see.

What was frustrating was that it was clear we had not really needed to turn back. With our powerful receiver I was sure we could have gone farther: just kept on, maybe for days, aloft with one of these radio-tagged falcons.

But neither Janis nor her tightly managed military program would ever make that choice. Chase had been assigned only to determine what percentage of the arctic peregrines, *Falco peregrinus tundrius,* that migrated up from the Tropics every spring then traveled northeast from the Gulf Coast. Others went west, maybe to Alaska. As only a falcon-trapper's helper, with no role at all in Chase's study, I was lucky to have maneuvered my way onto even a single radio-tracking flight, and though I desperately wanted to go up again, Janis's program already had its data. She was scheduled to move on to another assignment, and today was her last mission.

Maybe, I offered, I could keep flying with Vose — a long, silver-topped column of a man people took for Janis's dad — filing more logbook statistics. But Janis made it clear that after she left Padre the Army wasn't leaving its sophisticated radio-tracking gear with anyone. Especially, I gathered, either George or me.

Though the military had rented Vose's little plane, to the project's buttoned-down young administrators George hardly could have offered a less compatible job profile. As a World War II combat flight instructor he was a generation older than even the parents of his Army supervisors, and he'd logged more wartime military years and on-the-edge light-plane aviation hours than all his bosses put together. But beyond his occasional mention of his barnstorming years, none of them knew anything about Vose's past since they'd hired him principally because he was the only telemetry-experienced pilot willing to accept the low-paying charter Janis had to offer.

George's penchant for talking freely about his work also meant he was seen almost as a security risk, and I'd heard among the close-cropped young officers who ran the Army's program that it was necessary to keep an eye on him, though I couldn't imagine what kind of security breach his tracing the migration route of endangered birds of prey might entail.

I didn't rate even that much attention. A lifelong naturalist, birder, and writer of herpetology texts, I was no more than the friend of one of the program's directors, Kenton Riddle of the University of Texas's Bastrop Science Center. I had managed to hitch a ride on the radio-tracking plane only because storms had flooded the mud-flats where we trapped the study's peregrines. But after that first flight I couldn't get the falcons' journeys out of my mind, and when Chase left Texas two days later I saw her onto the flight back to Patuxent, Maryland, then drove over to Laguna Vista Airfield. Vose was stitching up the sagging cloth of his Cessna's headliner.

"Customs men," he grumbled. "*United States* Customs. Slipped out while I was doing their paperwork and cut up my plane. Looking for drugs."

He poked his needle into one side of a long slit.

"Wouldn't put it past 'em not to make sure they found some, too."

I examined the Army's yard-long, Christmas-tree–shaped antennas, which George had cobbled onto the Skyhawk's wing struts using hollowed-out chunks of 2 x 4 pine and a snarl of radiator-hose clamps. A little scary, but with no pilot's li-cense — never having even flown a plane — I wasn't in a position to be choosy.

"Ever think about keeping on?" I asked him. "Just stay up there with one of these falcons?"

Vose said he'd thought about it. He had even followed some, over longer distances with earlier researchers, and with Janis. He jerked his thumb at the Cessna's backseat, where three Army transmitters lay swaddled in bubble wrap.

"Still got some radios. . . . But what I'm saying is, this little Army deal's nothing. Not even real aviation."

It took Vose more than a moment to swing his big frame down from the plane, but he wasn't finished.

"Lemme tell you. . . . After the War, one of the things I did for a living was go in to crash sites. Just one helper. Build us a board runway, eighty yards long. Then we'd mechanic that aircraft back together and fly it out."

He looked at me closely.

"*Certainly* I could keep on after one of these little things."

I raised my brow.

"Clear across the country?"

George put down his needle and fishing-line thread.

"Any country in the world — falcons only average fifty or sixty miles an hour."

Anywhere in the world. I took a breath. Vose couldn't know it, but since our first flight, the idea of following one of these

18

creatures, wherever its airborne life might lead, had become the grandest idea I'd ever had.

George shook his head.

"Military won't approve it," he said. "They're waiting for a satellite — due out sometime in the nineties — that can track these little transmitters." I pictured an electronic warren, glowing with data screens where some technician sat, coolly plotting the intercontinental flights of flesh-and-blood peregrines as they streaked from tropic jungle to arctic steppe. By then, Vose and I and his little Cessna would be irrelevant.

Yet for a while now we still had a chance to be of consequence. If what he'd said was true, we really could go up with a peregrine, fly with it, and, while mystery still cloaked its realm aloft, share the primeval momentum of its dream, or instinct, or simple fancy.

I just didn't know how much any of that might mean to Vose.

He was an aviator, not a bird guy, so maybe not much. But he had definitely been disinclined to let that first peregrine go. That and not having to watch from the sidelines as his share of aviation dwindled into technological irrelevance might be enough to draw him into what was becoming my own overwhelming vision.

In any case, it was now or never. I took another breath.

"How about . . . well, why not try it on our own?"

George put his sewing gear back in the toolbox.

"These birds are protected; some of 'em endangered. Army'd never give us permission. . . ."

I looked dead at him.

"Of course they wouldn't."

Vose turned away, and I was sure I'd lost him. I had barely met him, and there was absolutely no reason for him to defy his military employers in favor of a semi-illicit trip with me.

For a while he studied the Skyhawk's stitched ceiling, which now resembled Frankenstein's forehead. Then, in the same careful way, he looked me over, stroking his mustache. It was thin, white, and pointy — Errol Flynn style — a holdover from his adventure-flying days.

"Aviation takes intestinal fortitude, Mister. You were pretty green up there today. Calm air, too."

From a battered leather suitcase George dug a roll of orange surveyor's ribbon.

"I don't do this," he explained, draping the pointed prongs of the Army antennas with fluorescent streamers, "somebody'll kill themselves and sue me." He looked over. "Can't think why you'd want to follow a falcon in the first place."

Want to? Follow a peregrine, maybe all the way to its polar breeding ground? No one had ever done such a thing. But I was talking to the one man in the world who could.

I caught George's eye.

"Same reason you'd want to," I said.

Above the Skyhawk, big white cumulus were building, and to avoid each other's eyes we both looked up at them.

"Go," I said, "where no one's ever gone."

"Ever known *how* to go," growled Vose.

But I could see his doubts. I was some kind of bird nut: no idea what kind of down-and-dirty flying my idea called for. Still, paying jobs in aviation didn't come often to a guy in his sixties. Late sixties. Bad knees, shaky hands. Gin and tonic more often than every now and then. Plus, this Army deal wasn't just discouraging; to a real flier it was almost an insult.

"We'll never get this chance again," I said.

Vose scuffed a foot.

Just barely, I could see a crinkle crease the corners of his pale blue eyes. He stepped back from his ribbon-draped antennas, stowed the roll of tape, and rooted through his leather case. Finally he pulled out a black plastic garbage bag.

"Here," he said, pointing at his throat. "Buy you some of these. Just in case."

At my grin he darkened.

"Now we're only doin' this, Mister, no more'n a couple days. Tops."

2.

Padre

Forty-eight hours later, gassed and as ready as its rattling old motor was ever going to be, Skyhawk '469 sat by the flight shack at Cameron County Airfield. I already knew my new partner would be inside swapping stories with anybody who even looked like they might have once flown a plane, while he waited for me to call and report that I'd captured a peregrine falcon and put one of our leftover Army transmitters on it.

Not that I was very likely to. I was camped at the far end of the largest barrier island, smack in the path of the tundra falcons' migration route, but so far I'd mostly just watched as Ken's trappers captured peregrines. I had studied the team's equipment, though, and since I'd hooked up with Vose I'd made my own rough copies. But I still wasn't sure exactly what to do with them.

The biggest problem was that during three days alone on the flats I had seen nothing resembling a peregrine. Then, before first light on the fourth morning I smelled the mainland. I'd been on Padre long enough to know that meant an onshore wind was pushing out from the coast, ahead of a late-spring cold front. Soon, rain would force the shorebirds into the shelter of the reeds, and the falcons would press for an early kill while their prey was still exposed.

It was their way. Fierce wild spirits so briefly seen as to seem more apparitions than real birds of prey, peregrines hunt at the edges of the day, when their enormous, light-gathering pupils give them an advantage over the small-eyed wading birds unable to fly well before first light. Seldom relaxing like other hawks in idle floats across sunny summer clouds, Padre's peregrines hid in the diminished perspective of distance, crouched on the empty flats, picking out prey birds with the uncanny acuity of their binocular vision. Then, concealed in the dimness of dawn or dusk, with a rush they would come flaring in, cutting through a sudden tumult of waders flailing into the air, then forever shutting off for one of them the light of the open sky.

A time or two I'd seen that here, but

more often I had just glimpsed peregrines flickering like big spectral bats over the pre-dawn flats. It was a hunting ground not shown on any map. U.S. Geological Survey charts show the South Texas mainland to be paralleled, 10 to 20 miles out in the Gulf of Mexico, by the long fingers of Matagorda, Mustang, and North and South Padre Islands. But instead of the broad bay that the maps depict as separating this 300-mile arc of barrier dunes from the coast, there is actually land.

At least there usually is. Except during the highest tides, most of what occupies this space is — despite its being called the Laguna Madre — dry, sandy terrain: a level, featureless desert the size of Connecticut. Known as a wind-tidal flat since the direction of the wind mostly determines whether the plain is submerged, this vast coastal plateau lies no more than a foot above sea level; without the barrier island's ridge of intervening dunes, the Gulf's 3-foot surf would roll all the way to the mainland.

Now the flats were empty, yet this desert of sandy mud was an oasis, a critical migration stop for the white-breasted, gunmetal-backed adult tundra peregrines. Thousands of miles north of their better-known but endangered southern *anatum* relatives — the more sedentary, often non-migratory falcons that nest on the sky-scraper ledges of a score of eastern cities — the nesting aeries of arctic peregrines ring the polar rim of every northern continent. From this arctic realm, each autumn, tundra peregrines wing their way across North America, dropping onto the off-shore Gulf Coast islands to rest and feed.

A few remain there all winter, but most tundra peregrines, each on its own solitary route, fly south. Maybe the strongest, the most willful, or most powerfully genetically programmed — maybe simply the most curious of their race — some may go as far as Argentina. But when our Padre migrants left the island, for the most part they simply vanished. I'd seen peregrines in the Caribbean during winter, and a few had been reported crossing the Isthmus of Panama, but no one was sure whether they were the same individuals that left the arctic steppes of Canada and Alaska with the first autumn winds. All that was known was that each April, from somewhere to the south, a depleted population of tundra falcons reappeared on Padre Island. Where they had been, and why so few returned, was still a mystery.

★ ★ ★

Arriving with the first springtime waves of northbound shore- and songbirds, the falcons were always hungry. Feeding almost entirely on other birds taken in flight, peregrines hunt most successfully over barren ground; where there is even a scrim of brush a fugitive passerine or shorebird can, at the last moment, dive among the wiry stems where a peregrine will not follow. On Padre's stark flats falcons have the edge, however, and here they wait in ambush, watching from the tops of old drift fences or washed-up chunks of timber for the flurries of warblers, vireos, and flycatchers that flutter in from the sea. Exhausted by their two-day flight from Yucatán, these small creatures are desperate for the sheltering foliage and insects of the mainland. But first they must pass the gauntlet of the flats.

Only a few fall to the falcons, yet every small life that sets out across Padre's savage desert feels the terror of the unseen eyes that scan the plain's featureless horizon, and in the dimness of evening I too could feel the falcons' presence — an essence far more palpable, out on those miles of empty sand, than the abstract medical research of the University of Texas

27

M. D. Anderson Cancer Center, which was a partner with the Army in Ken Riddle's project.

The university's part of the enterprise centered on the crucial testimony peregrines carried within their bodies: traces of organochlorine pesticides and environmentally dispersed carcinogens that could affect my life as much as theirs. To read that chemical signature entailed drawing a droplet of their blood, though, and that meant following the hawks' own ferocious karma. It meant offering them the lives of prey.

Our victims were decoy pigeons, and as I groped through their coop before dawn each morning I was sorry that their soft lives were the only gift that could draw down a peregrine. In the darkness, a storm of frightened pigeon wings hid the six I came up with and, smoothing their feathers, I slipped each one into its carrying box, mounted on the back of my ATV. The pigeons were quiet and comfortable in their sheltered niches, but I knew I could not treat them with the falconer's traditional detachment.

For one thing, they were too similar to falcons. With their slender-winged, big-breasted profile, pigeons' bodies aren't put together much differently than peregrines'. Their likeness may be what draws the falcons, for the rapid-fire wingbeat of Columbiformes seems to trigger in birds of prey a flash of predatory recognition. But theirs is a pact forged cleanly, in the heat of flight and pursuit between near-equals, and I resolved to try to prevent my hobbled captives' death in the talons.

Squatting like a fat red toad, my Honda ATV idled throatily in the murky air outside the trappers' beach cabin. Our balloon-tired ATVs, which each of us had outfitted to sustain him for days out on the flats, were the only way to reach the peregrines, and in the darkness I skimmed up the shadowy ocean beach, skirting pale ribbons of foam that snaked inward across the sand.

Then the sky bloomed pink, expanding my miniature headlight world into a silvery panorama of waves and dunes that glistened through haze so dense it let me look directly into the rising sun, against whose coral ball a great blue heron lifted out of the surf and, like a pterodactyl, labored seaward. I waited for it to turn back, but the heron's slow cadence brought it only to the center of the sun, where it hung, growing smaller and smaller until it vanished into the light.

In the other direction were the flats. It took a while to lever the Honda up and then down the far side of the brush-covered ridge that backed the beach, but beyond the dunes was another world. A few miles to the west lay the Intracoastal Waterway, yet the flats' murky tidal cosm was so surreal in its emptiness — so spooky, even — that old wilderness hands sometimes came in for good after only a couple of days' hunting.

Mostly it was the loss of perspective. Faintly corrugated, the laguna's sandy verges melted into an absolutely level, fea-tureless plain that at the indeterminable horizon dissolved into a marbled sky, flat as a scenery backdrop. In that hallucina-tory vastness familiar definitions lost dis-tinction, and before midday I had become a Brobdingnagian intruder, rumbling across a shrunken topography where for miles nothing reached higher than my in-step.

Yet I was not alone. In the pastel dis-tance what looked like a sheet of waving paper proved to be the snowy wing of a cattle egret — still lofted, over its carcass, by the wind. I shivered: feathers unrum-pled, the egret still bled from the surgical incision that sliced its belly. Only a peregrine would scissor apart prey with such precision. As I spun in circles searching the haze with binoculars, a nearby slab of driftwood caught my eye. Beneath it were painted off-white spatters. Mutes, or droppings, they could have been left by a gull or cormorant, but as I pulled up to the timber I could see that, along the rough bark, shreds of down had lodged where a falcon had bent to scrape its bill.

Here, in the dimness of dawn, a peregrine had returned with prey. I ran my fingers down the wood. Among the feather scraps were spots of flame, bits of orange and white plumage, and beside the log I found their owner. Its untouched black head showed it was a Baltimore oriole — a male — but as I pulled it from the sunken girder's shadow I realized the oriole was weightless. Gutted. Below its throat nothing remained but its harlequined wings and tail and its apricot flanks that still glowed with the gaudy hues of the tropics.

Reflexively, I glanced up. The hazy sky held nothing, but as I followed the horizon's circle, with my 10 x 40s I glimpsed a fleck of movement. Then a dark shadow came rippling in across the sand. Suddenly, 15 feet above me the shadow's maker — a clear, perfect peregrine — was etched against the sky. On one buoyant wingbeat it pulled up, and for a slow-motion instant our eyes locked; then with a scream the falcon flared away on the wind.

Where would it go next? I did not yet know peregrines well enough. All I could see was the picture this one had etched, beginning — as it always did with falcons — with that huge, shining dark eye, bordered by a ring of yellow skin, broader toward the pearl-gray beak. From this bird's size — it was nearly half again as large as a male, or tiercel — I could tell she was a female, and from her brown-mottled breast, a youngster, what the trappers called a hatch-year bird.

Her coppery crown and golden cheeks and throat meant she was a young *tundrius*, now on her first springtime trip back to the Arctic. That made her a survivor — one of what was thought to be a minority of adolescents that, southbound in autumn, reach the Gulf Coast again the following year. She was also hungry, or she wouldn't have checked me out, poaching her fresh egret and the dried remains of her oriole.

Maybe hungry enough, I thought, to come back to her most recent kill; so with shaking hands I slipped one of the pigeons into its noose jacket, a leather vest sprigged with loops of slip-knotted 20-pound monofilament. At the slightest touch the nooses would cinch tight, snaring any talon that brushed their transparent tendrils, momentarily tying a falcon both to the pigeon and to an attached drag weight light enough not to pop the line when the falcon took off but heavy enough to limit how far it could fly with its prey.

Stashing the dead egret on my ATV, I set my pigeon in the center of the scarred sand that marked the egret's struggle, hit the starter, and rolled a quarter mile directly into the wind. Then I turned back to face the now-invisible pigeon.

With a mammal who would pick up your scent you'd avoid an air current like that, but with a falcon it was the only angle of approach because, aloft or alight, peregrines live in the wind. Across blustery tundra or continental coastline, the press of moving air sculpts their lives as surely as falling water shapes the form of millrace trout, and like swift-water fish, peregrines are always angled into the current. If my falcon returned to her kill she would be more likely to do so by traveling into the wind, facing me as she came in and continuing to do so while she fed. That way she could see me clearly, but I could approach her more easily because always in her mind would be the possibility of the ready escape she could make by gusting away downwind if I got too close.

That was the theory, anyway. From my vantage point on top of the Honda's cargo box, through my binoculars I could just make out the capped, sand-filled PVC cylinder that was my anchor drag, and for the rest of the day I gazed at that progressively blurring ashen dot. Maybe my passing youngster hadn't killed the egret after all; she could have just happened by after I'd flushed away its real owner, and with no memory of a kill she had no reason to come back. So just before sunset I scooped up my pigeon and set out across the algae mat.

It was even weirder than the mudflats. A lumpy carpet of blackened seaweed stranded when the baywater periodically swept out to sea, the algae was crusted with desiccated clams and marooned marine life, offering a vaguely biblical reminder that the endless plain around me could, at any moment, again become ocean.

Strangely, falcons loved the algae. Out

on its glutinous surface peregrines were so hard to see I finally realized that the dried seaweed's attraction was as camouflage. Against the laguna's pale sand a distant falcon stood out as a sharp black speck; on the algae its mottled plumage was part of the background.

Not far ahead, part of that dark background showed a twitch of movement. It was too low for a falcon — a probing shorebird, I guessed. Then, damn! I'd gotten too close. In a flash of wings a peregrine, maybe the female adolescent I'd seen earlier, detached itself from the algae and vanished into the dusk. Its departure was so sudden that I needed to salvage some evidence it had really been there at all, so I eased the Honda forward, studying the ground.

Next to a rick of leaves lay its kill: a purple gallinule, marked only with a crescent of scarlet blood ringing the base of its skull. It was instantly clear why there had been no profile: the falcon had been bent to her killing. But she had not had time to feed, so I picked up the gallinule and set another of my now-terrified pigeons in its place.

If the peregrine returned and was the adolescent I'd seen, at first she probably wouldn't hurt the pigeon because most first-year falcons have yet to learn the high-impact dives their elders are famous for. If the pigeon stayed on the ground it would seem to be a sick or injured bird — prey near which a peregrine might simply land, then stride warily up to. The pigeon's leather vest was good armor against that sort of attack, but after the falcon grabbed it I'd have to get in fast, before the hawk lowered her deadly beak.

I only wished the pigeon was in on the plan as, aware of its peril, it crouched like a stone against the seaweed, watching the distant flicker of falcon wings working back and forth in the growing gloom. The peregrine had seen me near her kill, but not realizing the gallinule was gone, she was coursing the area in search of it. The surrogate pigeon was obvious to her, but its extra plumage of leather and fishing line seemed to put her off, and just beyond the patch of sand where she'd taken the gallinule, she hovered for a moment and then set down. With the short, toed-in strides peregrines use on the ground, almost like a parrot she trotted back and forth, scrutinizing every grain of mud. Then she glared at the pigeon.

That was too much for it. It started to

creep back toward the distant shelter of the ATV, pricking the hawk's attention and gradually drawing upon itself the focus of her hunger. There was time for me to feel crummy, but not much, for the peregrine pushed off into the wind and, one wingtip brushing the ground, banked toward the pigeon. It froze, then flattened itself as she ripped by an inch overhead. As she swung back on a crosswind pass I saw that the first time wasn't a miss. She was trying to flush the pigeon into the air.

But her quarry, which knew instinctively that it would be most vulnerable on the wing, wouldn't budge. Playing possum was its only chance, for most peregrines are reluctant to strike a target on the ground, and the longer the pigeon stayed down the more likely the falcon was to be distracted by other prey.

But that wasn't figuring on the darkness. As it deepened, the peregrine fluttered to the ground beside my quivering pigeon, squirted out a dropping, and fluffed her feathers. Aware somehow that her tethered prey could not escape, she had decided to save it until morning. Neither the pigeon nor I could have stood the suspense of that, so to break the spell I edged a little closer on the ATV. More secure now in the growing gloom, the falcon bristled at my approach. Then she turned to the pigeon and, like some bush-jacketed sportsman posed with his trophy, clamped a proprietary foot onto its back.

That was what I'd been waiting for. As I hit the gas the falcon flared her neck feathers and leaped into the air, carrying the pigeon with both feet. She was a big bird, and when she reached the end of the tether cord her momentum sent the drag weight skipping over the seaweed. Flying into the wind she couldn't gather speed, though, and I was gaining so fast that with a scream she dropped her cargo.

Or tried to. As the pigeon fell, its nooses snagged one of the falcon's middle talons, dragging down her suddenly imprisoned leg and catalyzing a wing-flailing panic that lifted her 30 feet off the ground. Betting she couldn't keep the whole train aloft, I bailed off the Honda at a dead run and in seconds was directly beneath her.

That was a mistake. Getting so close before she had tired gave the peregrine an extra shot of adrenaline, and she cut back over my head. There, buoyed by the wind, she began to gain altitude. As the pigeon's drag bounced clear of the sand for the last time, I dove for it . . . and, unfortunately, connected. There wasn't even a shock at my end of the line as that single toe-holding strand of monofilament let go.

3.

From Forayne Partes

Sunburned and weary, I lay facedown in the rotting algae. Of course the line snapped. That's what it was supposed to do — slow a peregrine enough to eventually wear it out and bring it down, yet let the falcon break off easily if it got away from the trapper.

Like some pirate victim buried to the nostrils, I watched rows of wavelets, fore-runners of the incoming tide, advance toward my face. Then the pigeon's soggy wingbeats brought me upright, and I felt even more guilty. Except for being damp the pigeon was all right, though, so I put it back in the wooden cage on the ATV. By then a warm demi-surf had risen around my ankles, stifling the algae's benthic smell as all the water that the on-shore wind had formerly swept out into the Gulf came flooding back into the laguna.

Within an hour, a shallow sea stretched all the way to the mainland, and even by starlight I could see that for miles nothing protruded above the surface except me and my small machine. The Honda's carburetor had a snorkel that kept its motor running, so it became my small, mobile island, burbling along like a tugboat, churning a pale bow wave.

Then the moon rose and the sea's surface lit up so brightly that I shut off my headlamp and followed the silver path of light that, at the horizon's seam, stitched me to its lunar source. A Shining Path. Appropriated by Peru's revolutionary leftists, it was a name I'd heard constantly in the Andean highlands — one of the places I led nature tours — the site held by some ornithologists to be the enigmatic winter home of *tundrius*. The peregrines I saw there, though, on the mossy precipices above the Urubamba River, were resident birds of the local *cassini* race.

Wherever today's lost juvenile had spent the winter, during April Padre was only a momentary stopping point for her, because across its pastel desert the force that had brought her here reached on, perhaps 3,000 miles to the arctic cliff where she was born. It was a path without markers, an airborne passage whose coordinates lay plaited among the synapses of her brain,

41

inscribed there by generations of success and failure.

But what lay within her was more than merely a route. Partly learned and partly instinctual, the pathway this young falcon followed was only the smallest part of the mystery of her migration; most of its riddle lay in the enigmatic power that was, even now, driving her on, day after day. Toward the Far North, toward the tundra: Greenland, Nunavut, Alaska. It was an ancient yearning, separate and apart from the world of man, and realizing that in some way I had almost touched that force, I plowed my small wake on across the bay toward Deer Island, a knoll where there was dry ground to camp.

If there were ever any deer on Deer Island, during some of the nights I camped there they'd have been up to their bellies in storm tide. Above the waves only a few dozen square yards of dune remained: a cartoon desert island that lacked only a single leaning palm. On that little knoll I fed and watered the pigeons and crawled into my sleeping bag, flopped next to the Honda's fat rear tires for protection from the wind. Out of the Army's larder I'd drawn combat rations — cans of spaghetti, biscuits, and Spam — plus water and half a case of Gatorade jugs.

The Gatorade was the most important. Dehydration is a constant threat on Padre's desert of mud, and I ordinarily drank one of the quarts with my C-rats, the other later in the night. For easy access I wedged one of the big glass jars into the sand next to my head.

I don't know how long I slept, but suddenly something was wrong. A funny scent. I reached for the Gatorade, but there was only a depression in the sand, and with a shudder I recognized that musky smell: 18 inches from my nose a pair of coyote prints dented the ground.

I knew why they were there. The day before, trying to pick out a ground-perching hawk from the frieze of driftwood lining the skyline, I'd seen movement in a gray lump of flotsam. In the evanescence of the flats it could have been a nearby godwit or a far-off jackrabbit. But it was alive, and as I idled toward it the distant shape gradually resolved itself into a trotting coyote.

As it noticed me the little wolf broke into a lope, heading east toward the shelter of the barrier dunes. Yet in the soft sand he couldn't run well, and as I pulled up on him I could see his heaving flanks and hear his staccato pants and the scrunch of his paws on the crust. He seemed no more than a thin, scraggly dog, and I remembered hearing that the year before, a half-grown coyote had snatched one of our bait pigeons, whose veil of monofilament had snagged in its teeth. Had the harness then shaken loose, its pigeon-baited nooses adrift on the flats might have destroyed a peregrine, so after a 10-mile ATV chase the trapper had run down and killed the hapless coyote.

For all it knew, today's little wolf faced the same fate, yet 5 yards ahead of my oncoming machine, he made no desperate sprint: facing his poor set of options, he watched me over his shoulder and galloped resolutely on. Ashamed, I cut the throttle. But it was too late to turn away unconfronted. The coyote had made his choice and, with Zen-like resignation, turned to face me. As he gasped for breath, his pink tongue lolled from the side of his jaw, but his flaming yellow eyes, lit as if by yellow candles, were riveted on mine. For almost a minute our gazes remained locked; then I glanced away, and in that instant he was gone.

Digging deeper into my bag on Deer Island, I wondered if tonight's coyote could have been the one I chased, come back in the night to gnaw the heavy glass of the Gatorade jug. Dehydration was his enemy, too, since most of Padre's land animals live off the sea. They swallow salt with every bite of food, and in their stomachs those saline ions draw in less-salty liquids from their blood and tissues, leaving their bodies constantly on the brink of fluid starvation, which is why you can't survive by drinking even a little salt water.

Ocean-dwelling reptiles — like the Ridley's sea turtles that recently nested by the hundreds on these beaches — purge extra salt through specialized glands located in the corners of their eyes; shorebirds are sustained by the acceptably brackish body fluid of marine worms and crustaceans; and peregrines live off the blood of other birds. But the barrier island coyotes have no regular source of water, and I rolled over, rooting for the bold little canid who had waded ashore to crack the glass and lap my Gatorade.

Turning over let me see, through a notch in the island's low dune ridge, the orange skein of bulbs glowing over an offshore oil-drilling rig, anchored like a cruise ship out in the Gulf. But its carnival brightness already belonged to another world, far from the half-imagined realm these peregrines were leading me toward. It was a domain of noblemen, serfs, and sorcerers, an ancient kingdom where the falcons' annual arrival from the icy wastes beyond the known world so stirred the dreams of men tied always to their fields that they named this bird peregrine: the wanderer.*

Sometime before dawn the wind swung around to the north and calmed the waves enough for me to hear, drifting down from the quieted sky, faint peeps. They sounded ghostly and sad, but they were really cries of happiness, issued by excited songbirds coming in from the Gulf with the mainland now in sight. Among them were more orioles that a day or two ago, on the shores of Yucatán, might have fed on the same ripe mangoes as the peregrine-killed male I had held that afternoon.

A momentary morsel to its captor, the

*As early as the twelfth century, the literary term *peregrina*, later *peregrinate*, meant to wander or travel. *Falco peregrinus* was probably first used as a scientific name in 1250 by Albertus Magnus de Animal, although until the sixteenth century the term *peregrine* was often associated with any migrant bird of prey.

oriole had lived a life no less epic than that of the peregrine that snatched its bright Halloween colors from the sky. The day before it died it had fluttered, 400 miles to the south, among tropical foliage, plucking caterpillars from the leaves and probing fallen fruit for its sweet pulp and wriggling fly larvae. Then, sometime between noon and darkness, it had launched out over the waves. All night, high above the sea it had held its course, pumping out four wing- and heartbeats every second for sixteen or more hours. By morning, from nearly a mile up, it would have seen the thin brown line of the Texas coast.

Again this evening other orioles, traveling in waves that included grosbeaks, tanagers, flycatchers, thrushes, and more than a dozen species of warblers, would gather their collective force of will, there on the northwest shore of Mexico's coastal horn. Carrying their race's genetic hope for the future, they would leap onto their fragile wings, eager to surmount the void of wind and wave and vast distance that lay between them and their breeding territories a continent away to the north.

In the still air, most of tonight's voyagers had made that crossing. But with no way to anticipate what lay ahead, tomorrow's flocks might not. Two or three times every spring a mass of cold air surges down from the north and meets the migrants partway across the Gulf. If they hit its storm-ridden headwinds near the Mexican coast, almost all the birds turn back to try again another night. But those who have passed the halfway point, radar has shown, choose to continue, fighting the wind. Some of the strongest, like the kingbirds, usually knife on through. But among the lightest and most delicate pilgrims — buntings, wood-peewees, weak-winged thrushes, and especially the tiny warblers, vireos, and hummingbirds — thousands fail to make their Texas landfall.

For the most part they don't falter because of exhaustion; it is a lack of fuel. Birds' fuel is not food held in their stomachs, which are minimal, or even in their crops, which are larger. Birds' fuel is fat. Unlike that of mammals, avian muscle isn't laced with lipids; birds' gas tanks are the flat pockets of yellow suet deposited along their breastbones, in the lateral hollows of their necks, and in their wing pits. When these supplies are exhausted, individuals flying in laboratory wind tunnels don't gradually lose muscle strength from lactic acid buildup, like an exercising mammal;

when they are truly out of fuel birds just come to a halt and quit flapping altogether.

Because the stakes are so high, songbirds migrating over open water can continue for a desperate while after their fat is gone by consuming their own bodies. But the only fodder their bodies have is flight muscle, and cannibalizing that brings them down to their last hope: the boundary layer of air just above the water's surface that offers a bit of added lift. Riding it, many migrants fight on the last few miles to land, but if the north wind remains strong, each spring huge numbers of songbirds fall unseen into the waves. Then thousands of small bodies — a tiny fraction of those who perished in deeper water — wash ashore with the morning tides.

I'd seen those urgent comings-in.

Like some sort of alien bee, a ruby-throated hummingbird had whirred past the offshore boat on which I was working as a teen and cut a quick 180-degree turn. Three feet off the water, barely holding himself aloft on boundary-layer buoyancy, the hummer made it back to our rigging, where he rode most of the day as we chugged slowly toward land.

Farther up the coast, at High Island, the

same springtime comings-ashore happen daily. Unlike Padre, this is not really an island but a chenier, an almost imperceptibly elevated little plateau of ancient riparian deposits that bulges up just enough from the surrounding saline marshes to let copses of live oak and magnolia, greenbriar, locust, and prickly-ash dig in their roots.

In good times, most songbirds have enough flight capacity to continue on, high above the inhospitable coastal marshes, to descend, 50 to 100 miles inland, among the canopy of East Texas's tall hardwoods. But when northerly winds press small passerines down onto the land's salty edge, High Island's cheniers save millions of songbird lives.

But the birds that reach those diminutive shoreline forests are changed creatures. Wariness is a lost priority: all that matters is food. In and out of every trailside bush, no more than two or three arms' reaches away, I watched chestnut-sided, sky-blue cerulean, orange-bellied parula, and oriole-colored blackburnian warblers, bold as they never would be in the high-tiered northern forest where they breed. Above the warblers, a brace of summer tanagers had taken turns picking off bees emerging from their tree-crevice hive, and a notch higher, rose-breasted grosbeaks — sparrow-like females as well as medallion-chested males — pulled off every reachable mulberry, ripe or not.

For birds unable to reach the cheniers, even making landfall could be disastrous. After fighting an all-night headwind, flocks of indigo buntings had fluttered onto the beach, the weakest ones plopping down right beyond the surf — some on the sand, more on the sparse lawns of beach cottages. Most had enough left in their wings to soldier on a few more yards to the barbed-wire fences bordering Texas Route 87, where, after a minute's rest, looking like bright confetti, wave after wave of them set off across that 20 feet of black asphalt.

Heedless in their need to reach the insects of the grassy fields on the other side of the road, a yard above the ground the survivors of that long ocean crossing set out — only to be met, sometimes five or six at a time, by the 70-m.p.h. rush of spring-break-vacationer traffic. From the roadside, where hundreds of brave migrants lay crushed, I'd screamed at every passing vehicle to slow down until — ignored as a madman from behind air-conditioned glass — I finally accepted the futility of trying to intervene.

Storms have always taken a heavy toll on these voyagers, and passerines' high reproductive rates historically managed to offset the losses. But that was before the coming of mankind, before the human glacier that now bears down on every continent in the world. To an animal, of course, death is just death: it doesn't make much difference if its end comes in the talons of a peregrine or on the grille of an SUV.

But overall it matters. It matters when technology obliterates not just the life but the heroism of other species, takes all the heart that has carried tens of thousands of generations of buntings or warblers across a vast dark sea to the far springtime shores from which they can find their way home — and smacks all that soul and immense bravery into oblivion. Falcon kills don't do that. The swift raptors that patrol the borders of the migrant stream are *part* of the stream, and the relative handful of birds they devour — sometimes the weak and sick, but usually just the unlucky — nevertheless still go on, the nutrients of their bodies fueling their predators' part of the same vernal tide that they, with such enthusiasm and trust, committed their lives to joining.

★ ★ ★

At midday, 10 miles up the beach I found a juvenile ring-billed gull. It was surrounded by the signs of its death, among them the tracks of a peregrine. The falcon's forward toes were long, and its rear talons had left deep claw-marks in the sand, evidence of the curved black nails that, powered by the momentum of a dive, are the scimitars peregrines use only on large prey. Unlike the chewed-apart carcasses left by most predators, falcon kills were always cut open cleanly, with the breasts and viscera meticulously devoured, but because even middle-size prey birds are difficult for adolescent peregrines to overcome, this gull was more likely the victim of an adult, probably a female. Full of breast meat, she'd be impossible to trap, so I threaded my way back through the dunes.

They were my favorite part of Padre. Between the sand hills' pallid walls, narrow valleys sheltered a garden of tenacious desert plants. Bitter panicum, sea purslane, and gallardias spiked up between the ubiquitous sea oats, gulf-dune paspalum, and goat's foot morning glory. All were working to improve their joint survival, steadily knitting the dunes' sandy crowns into denser soil.

The plants' enemy was hurricanes: every few years one flattens the ridges, then rebuilds them as barren drifts. But between big storms the roots advance. It had been four years since the last major gale, and below the dune-capping thatch enough moisture had collected to grow dwarf sundews, whose plush rosettes were fatal flames to insects snared by the transparent tendrils that sprig their leaves. The sundews' presence indicated nitrogen-depleted soil, the same sort of substrate that in mainland bogs has also spawned carnivorous plants that appropriate their nitrogen from living prey. Around the sundews' nibbled edges I found the prints of a ground squirrel and the curvilinear track of a diamondbacked rattlesnake (the only serpent here whose belly path would be so wide) that hunted it, with both sets of tracks overlain by the crystalline scratch marks of the ghost crab, *Ocypode*.

Beyond the small herbarium of the sand hills, the flats glowed in the morning sun. For once there was no haze, and as I swung my binoculars across their golden expanse, in microscopic profile stood a far-off peregrine: another female, but darker than the pale tundra hawk I'd lost the night before. Her color likely meant she belonged to a dusky woodland race, nesting perhaps among the dim conifers of Canada's boreal forest. But having watched me flounder in and out of the dune canyons she clearly wasn't wary, so in full view I motored right out onto the flats, angling away from her in hopes that my departure would be construed as non-threatening. Right hand on the throttle, I reached back for a noose-jacketed pigeon and, without slowing, tossed it as high as I could.

Glancing back, I saw the pigeon realize it was in the air and start to flap, even gain a little altitude. That was what I wanted: a flight long enough to catch the peregrine's attention. Then all I knew was sand: that my mouth and eyes were packed with it and that I couldn't breathe. Digging clots of dirt from my tongue, I realized I was up-side down.

Why wasn't I on the ATV? As the air came back into my lungs I sat up, smelling gasoline. Above me I could see the Honda, pitched over onto its cargo box and leaking fuel from its sprung gas cap, so I squirmed away, showing myself in the process that I wasn't badly hurt.

As I got to my feet what had happened was obvious. From its downhill side the little sand-hill scarp stood out sharply. It was a barchan dune, swept into its characteristic mainland-facing crescent by the Gulf wind. An evening sun would have painted shadows below its 3-foot drop-off, but the direct overhead light had erased the demarcation of its lip; without seeing it, I had driven off the edge.

Ten feet beyond the wheels-to-the-sky Honda, lying as neatly as if I'd set them on the ground, were my glasses. I slipped them on and limped over to the machine. Despite its twisted bars and gouged gas tank it looked rideable, and as I bent to heave it onto its tires, a flash of wings caught my eye.

Beneath my shirt the binoculars were still around my neck. As I lifted them to check on the abandoned pigeon, there, beating heavily with the burden of its lifeless body, was the dark falcon. I'd dropped out of sight when I flipped off the dune, bringing her in to the pigeon, where she'd snared herself. As my drag skidded along the sand beneath her straining wings, the falcon's airspeed fell, and I knew I had her.

Fifty yards back, I climbed off the ATV and shakily trotted in. Wings and tail spread wide, the peregrine lay like an intricate feather tapestry spread against the sand, the chocolate-gray mottling of each dorsal feather rimmed with a saffron line that, joining others, spread a golden net across her nape and shoulders.

Closer, though, all I could see were her eyes. Huge and unworldly, they stared up like those of some small, ferocious angel, astonished at her sudden inability to spire away. Vision was her armor, her strength against the world, and the force burning from her face so transfixed me that, oblivious, I reached to touch her. Affronted at my advance, the peregrine wrenched onto her back and with a reptilian hiss snatched my hand in her untethered foot. She had not taken her eyes from my face, and her speed and accuracy at seizing a different target were so astonishing it took a second for me to realize she'd pierced my hand.

Settle down now, I said aloud, snuggling her torso between my elbow and side and easing the arched black ice picks of her talons out of my thumb and palm. Then I aimed both clutching feet away from my body and wrapped her yellow shanks together in a band of adhesive tape. That let me smooth her plumage and slip an elastic stocking down over her body.

As the peregrine felt her wings restrained she shivered but kept her vivid black-and-white face fixed on mine, reading every flicker of expression and opening her beak to hiss if even for a moment I stared back at her. I fumbled with the old falconry hood I'd borrowed, then as I got it over her head we both calmed down.

Now it would be easy. With nail clippers I snipped away my tangled monofilament nooses, spread an aluminum Fish and Wildlife band — 987-7717. WRITE WASHINGTON, D.C., U.S.A. — and snicked it around her slender saffron leg with hollow-nose pliers.

Next I unpacked the transmitter. There was no hurry, and with care I slipped the tiny radio out of its packing, finding it was about the diameter of a pencil eraser, with an epoxy head connected to a wire-thread tail antenna that made it look like a giant sperm. A magnet taped alongside its sending unit stifled the tiny hearing-aid battery, saving its limited energy until the last minute.

Free of the magnet, the transmitter's encapsulated microcircuitry would send out at least four months' worth of electric pulses — 1644.438 on the receiver if the military had gotten their frequencies right. I laid the transmitter next to the shaft of one of the peregrine's two central retrices.

58

Even in a full flare this medial pair of tail feathers never spreads, and a miniature sending unit attached to these quills has the best chance of remaining in place. I marked points where the transmitter's four pinpoint mounting holes lined up, and with surgical linen nooses cinched them to the base shafts of those big feathers. Each noose joint got a drop of cyanoacetate adhesive, which I also used to glue the antenna wire down the center of the falcon's right tail feather; when she preened she'd brush over the soft metal filament with no more attention than if it were a slightly frayed quill.

It was time for her to go. As I lifted her hood, the peregrine clicked her beak like an owl and lunged for my eyes. But just having emerged from the dark, she missed. Holding her farther away, I eased the elastic stocking down over her legs and tail onto my hand, which still gripped her clenched toes, and removed the adhesive tape. Then, trying to remember everything, I paused to admire her. She was the darkest peregrine I had ever seen, heavily barred below, with a uniformly black cap and nape, unlike the pale-cheeked, glaucous-backed tundra birds I was used to seeing here. I could have studied her all day, but I took a breath and stepped backward into the wind, setting her lightly out onto its current.

Looking back, the falcon's eyes locked on my face for the first 10 feet. Then she flicked her wings and flowed away, sliding 100 yards downwind on the first strong gust. I expected her to rocket out of sight, but instead she cupped her flight feathers and hovered back onto the sand.

Astonished, I realized this was the boldness that warrior societies and medieval kingdoms had found so inspiring. Now that she was free, to this dark peregrine I was again no more than some meager, lumbering thing, helpless to touch the swift power suddenly returned to her wings. What was important now was to preen — to audit every detail of her precious, complicated plumage — and glaring back, she deliberately bent her black-tipped bill to comb away the horror of my touch.

I wanted to watch every detail of her aristocratic grooming, but the Honda was still running, and if the oncoming norther held this falcon on the island long enough for me to reach the plane, Vose and I might have our chance.

Stung by the first pellets of rain, crab-like I scrabbled the ATV — none of whose wheels remained in alignment — across the flats toward the beach. Deer Island was right on the way or I'd have abandoned my gear to save time, but as I came up to the big red gas cans that marked my camp, a pale lump appeared between this morning's tire tracks. It was the Gatorade, its label gnawed to shreds and its orange lid pocked with tooth marks. The cap wasn't punctured, though, so I cranked it off, guzzled down the whole tepid quart, and, with the storm boiling in behind me, headed for home.

From the first Texaco on the outskirts of town, I tried the airstrip number. It just rang. I was afraid of that. I'd been out for days, and George might not have been as committed as he seemed. His Army bosses could have gotten wind of our plan. Or he might have just changed his mind and gone home.

But an hour later I saw that the old brown Chevy that Cameron's flight office had lent George for being such a good fuel customer sat at the end of the Motel Del Rio parking lot. At least Vose hadn't left, but as I banged into the room he looked up from his half-packed suitcase and my heart fell.

"Radio's working," he grinned. "Just picked up our transmitter over the laguna."

He squeezed his bag shut.

"You ready to fly?"

4.

High Romance

If ever an ancient chivalric code governed the lives of a bunch of ostensibly dispassionate contemporary scientists, it was the lure that falconry held for Riddle's peregrine trappers. In contrast to the Army crew, every one of Ken's guys was a fanatic for birds of prey, treasuring his collection of books on hawking, ancient and modern, and maintaining at least one hunting falcon in a specialized mews — a coop large enough to house a compact car, with a full-width, barred front window.

Besides their libraries and their peregrines, what united them was that, like Ken, every member of his team had done whatever it took — stacking vacation time and sick leave, pressing the usefulness of their varied backgrounds on their universities' funding committees — to get to Padre. Their résumés referenced a dozen far-flung disciplines, but in their hearts every one of Riddle's men saw himself most fundamentally as a falconer.

Ken's fancy was snared in junior high study hall. Flipping through a 1920 *National Geographic*, he found Louis Agassiz Fuertes's spectacular paintings of sheiks on white stallions pursuing antelope with teams of hunting hawks, of a gyrfalcon blasting a heron out of the air, and of beautiful peregrines leaping from the gloves of galloping medieval lords and side-saddle-mounted ladies.

It was a vision to stir a boy's blood, and the next spring Ken captured his first bird of prey. For a while he paraded around with the partly tamed young redtail perched on his wrist, but pretending to be Henry VIII soon wore thin. Then Riddle crossed the watershed and fell under the spell of the hawking life that for centuries had been the beloved pursuit of every outdoor man not tied to a plow. A more demanding prairie falcon took the redtail's place, and with it Riddle found his daemon, which was launching the mortal aerial chases by which falcons live. Yet, as it does for every falconer, immersion in that fleet-winged ballet — a spectacle that had fired men's souls from Britain to Cathay — came at a heavy cost. Hunting with birds of prey is an emotionally costly endeavor because with every flight comes the risk of seeing the hundreds of psychically draining hours of behavior modification lavished on a coursing hawk simply disappear as it flies away. For birds of prey are not dogs, who value their owner's company as much as his food and shelter; even the most docile raptor can in an instant slough off the tenuous bond that ties it to its handler and never come back. That happened during one early Air Force vs. Wyoming half-time show, when the Air Force Falcons' most meticulously conditioned prairie falcon circled high enough above Laramie Stadium, got a look at the plains beyond and, in a flash of wings, was gone.

Thirty years ago few American falconers would take that risk, flying their hunting hawks only close to the ground in brief pursuit of leather lures or pen-raised quail. But by the time Ken entered veterinary school in Colorado Springs, he was flying falcons like a medieval prince.

Sending his peregrine into the 10,000-foot thermals kicked up by Pike's Peak, he would jog across the plains, watching his bird rise out of sight beyond the clouds. If all went well, by the time Ken flushed some hidden grouse or pheasant his falcon would still be up there, hidden in its height, waiting to come barreling down into whatever prey had leapt, unknowing, into the air. Sometimes, though, that world beyond the clouds would be too large and Riddle's peregrine would disappear. Then, in a cycle every falconer knows too well, a new bird would have to be trapped and the months of training begun again.

But there was one more level — an emotional plane it took Riddle years to reach.

"It doesn't matter to me, now, whether my bird kills anything or not," he told me one night at the Padre beach camp. "What I'm after is the perfect flight. Seeing my falcon — any falcon, really — go up against the fastest, most agile prey. It takes place in nature anyway. But by setting it up, you know, I get to watch."

I knew. Years before, laid back in a canoe drifting through the Rio Grande's Big Bend canyons after my river-rafting clients had gone home, I'd seen three blue-winged teal jump out of the water, head downstream, then reverse themselves in a startling about-face. A second later I saw why. From the rim of Burro Bluff, 1,000 feet above, a speck had detached itself and, doubling its size every instant, was falling like a meteor between the canyon walls.

Clouded by its speed, the falcon reached the level of the terror-stricken teal with velocity too high to impact one without killing both the duck and itself. But that wasn't its plan. Fifty feet behind the teal, with a flick of its wings it wrenched out of its dive into level flight 50 miles an hour faster than the ducks' top speed. They were right above me as the peregrine came up on them, and beneath the hawk's rippling flanks I could see it unsheathe its hidden yellow feet and swing them forward, dropping its long hind-talons that opened like a Barlow knife to rake a silent explosion of feathers from the last teal's back.

A second later, predator and prey were 200 feet apart, the peregrine veering skyward, the lifeless teal smacking into the muddy bank behind my boat. I never saw the falcon return for its kill, but as I rounded a downstream bend, drifting flecks of down still floated above the water.

Dives like that are what falconers live for, but even without them almost everyone who sees a peregrine is smitten. Up close, what draws you is their eyes. Cloaked in a carrying hood that shields the bird from the terrors of abrupt human activity, *Falco peregrinus* seems no more than a slender brown chicken with bad-looking feet. Unhooded, its bottomless brown orbs, set beneath a miniature eagle's brow, radiate nobility.

To humans, that is. But raptors' eyes were never meant for us to admire. Weighing more than an ounce for a large female peregrine, their enormous lenses make pin-eyed shrews of people, for whom eyes of comparable size would weigh more than 4 pounds.

Partly because of their size, peregrine eyes do things we can scarcely imagine. Evolved to gather every photon of light at dusk and dawn — that time when for humankind colors fade and outlines blur — they provide binocular vision over some 130 degrees of forward sight, the only area employed in high-speed chases. In order to look elsewhere, a falcon, which lacks the rolling eyes of mammals and has instead a limited ability to move its eyes within their sockets, ceaselessly scans its surroundings with slight, abrupt movements of its head. Then, by means of the million or so image-sharpening visual cells present in every square millimeter of its retinal fovea, it is able to magnify any point in its field of

view into an image at least eight times —
and maybe twice that — clearer than any
human eye will ever see.

To achieve such resolution, peregrines'
ocular globes bulge outward from the
skull, a prominence that gives them the no-
ble-seeming brow they share with many
raptors. Cored by a bony shelf called a
superciliary ridge and capped with a dense
ruff of stiff little feathers, a falcon's natural
visor both shades its eyes and shields their
vulnerable protrusion. In the same way, the
dark cheekbone eye-liner humans find so
striking is a widespread predatory adaptation that in birds of prey camouflages their
bright, predacious stare and deadens the
glare of water-reflected sunlight.

Forward of that mascara, centered in the
yellow cere above the beak, are small
keratinous tubercles that fill most of the
opening of each peregrine's nostril. The
little bumps look like they'd impede
breathing, but instead they baffle the velocity of wind driven in so forcefully
during high-speed dives that it would otherwise balloon the falcon's lungs. The
compact beak, proportionately shorter
than that of other hawks, provides an unusual amount of muscle attachment — a
necessity for powering the tomial cutting-

edge points found mainly on the upper mandible. These subtle notches, often too minimally defined to easily catch the eye, are nevertheless important. They are the falcon's fangs.

Soaring hawks and eagles kill with their powerful, dexterous toes, driving their talons through the rib cages of their prey. Against big prey, peregrines use their beaks. In the air, however, some peregrines, like my Rio Grande falcon, learn to use their long hind-talons like a pair of rapiers. But most of a falcon's killing takes place on the ground. There, large quarry is dispatched by bending over a pinned-down captive to carefully work the beak's sharp tomial points between the victim's cervical vertebrae in order to sever its spinal cord.*

*The mechanics of this process were familiar to Native Americans. The Ohio-Hopewell, a group who occupied the Mississippi Valley sixty generations before Columbus, held a falcon deity to be their supreme warlord. Sculpted in copper mortuary statuettes, this raptor-headed combatant brandished a mace in one claw and a human skull in the other, but the real source of his power was his hugely exaggerated tominal teeth.

Like other creatures culturally vested with intense human emotion, peregrines have been among the first species to fall before contemporary technology. Their decline, whose cause took decades to decipher, was caused not by hunting or even by habitat destruction but by humanity's inadvertent chemical assault on the environment.

Shortly after World War II, European peregrines started to disappear. No obvious reason was evident, but even in places where they had nested since the Middle Ages, fewer birds returned each year to the old aeries. In the same mysterious fashion, across the Atlantic deformed or stillborn peregrine nestlings were noted in places where the habitat appeared to have gone unchanged for hundreds of years. This occurred along the eastern seaboard, in Pennsylvania, and in faraway California, but at that time chemical poisoning was thought to be a temporary, local phenomenon, if it was considered at all in connection with wildlife. In an era of nuclear detonations in the Nevada desert, even the scientific community failed to suspect the continent-wide environmental effects that were later found to be generated by poisons as seemingly mild as the insect-killing pesticides that humans some-times breathed all day without immediate, obvious harm.

In Europe, however, peregrines had been so revered in literary as well as sporting tradition that their decline was noted with greater concern, and in a search for the reason so many of the abandoned aeries were still surrounded by undiminished flocks of prey birds, Derek Ratcliffe of the U.K. government's Nature Conservancy began to climb to some of the old nest scrapes. All that remained were a few shell fragments from the sites' last eggs, but Ratcliffe noticed that they seemed flimsier than he remembered.

That finely tuned perception at first seemed to be impossible to verify — until Ratcliffe remembered the egg collectors. Throughout Britain, scores of blown-out peregrine eggs, some dating back to the seventeenth century, lay preserved in pri-vate collections, and after months spent gaining access to these hidden treasures, Ratcliffe determined that the shells of post-war eggs were, in fact, up to 20 per-cent thinner than those of earlier samples. That would have made them too fragile to have supported the weight of their brooding mothers, so many of the chicks developing inside the shells were crushed before they hatched.

The year was 1947, twenty-three months after dichlorodiphenyltrichloroethane (DDT) went into worldwide circulation. At the time, no one made the connection, but over the next twenty years this corrosive poison continued to eat away the reproductive capacity of many birds of prey, especially that of peregrines. Writing of Britain in 1966, J. A. Baker observed:

> Few peregrines are left, there will be fewer, they may not survive. Many die on their backs, clutching insanely at the sky in their last convulsions, withered and burnt away by the filthy, insidious pollen of farm chemicals.

At about the same time, though the native *anatum* peregrines had been gone from the eastern United States for more than a decade, the University of Wisconsin hosted the first international conference on *Falco peregrinus,* after which Cornell University established the Eastern Peregrine Falcon Reintroduction Program. By then the laboratory evidence was in: DDT was found to pass from insects to the shore- and songbirds on which peregrines feed. Concentrated by each successive transfer up the food chain from insects through insectivorous birds to bird-eating falcons, DDT's derivative, dichlorodiphenyldichloroethylene (DDE), could reach as much as a fortyfold concentration in the most heavily poisoned peregrines. Although even at this level of toxicity few peregrines were killed outright, east of the Mississippi the entire population stopped rearing young because even in small doses DDE so suppressed the transfer of calcium from the wing bones of female falcons to their developing eggs that few of the shells were strong enough to support the weight of an incubating parent.

Then the same thing happened on the West Coast. By 1970 only two known pairs of peregrines still nested in California, where twenty-five years earlier there had been 200. When the Environmental Protection Agency banned DDT in 1972, Tom Cade — who, as director of Cornell University's Peregrine Fund, was almost alone in believing such a thing could be done — took on the monumental task of reintroducing peregrines to the eastern states. Seeding the old nest ledges with captive-raised young bred from both the arctic *tundrius* subspecies and the indigenous *anatum* race of peregrines still found, in places, in the western states, Cade succeeded. Within ten years he could point to scores of breeding peregrine pairs nesting both on their ancestral bluffs and, often as locally cherished celebrities, on the decorative ledges of man-made skyscrapers throughout the Northeast.

Meanwhile, the last large population of native peregrines left in North America — the arctic *tundrius* — was also coming under study. Every spring and autumn, arctic peregrines, most of them thought to be from Greenland's nesting population, appeared on North Carolina's Outer Banks, where trapping them to monitor their body chemistry started. Other tundra falcons had shown up on the Texas coast, and though their numbers were few, geography intimated that many of the *tundrius* that moved south from Canada and Alaska every fall should pass down the Gulf shoreline.

Getting research funds committed to such a tentative supposition was difficult, but like Cade's one-man show at Cornell, Texas's peregrine project was fueled by the determination of a solitary, falcon-possessed scientist. One first inspired by *National Geographic*.

Battered, muscular, and invariably clad in jeans and running shoes even in his director's office at the veterinary division of the University of Texas M. D. Anderson Cancer Center, Riddle had come, by his late forties, to resemble an even craggier version of Chuck Norris. His graduate work was with primates, but since that long-ago adolescent study hall perhaps the most important part of Ken's life had revolved around birds of prey, and in order to spend two months of every year in the field with peregrines he had merged an elaborate inquiry into their metabolism with the university's cancer research.

That had taken some convincing of the university's top-level administrative suits, because to gather enough data for meaningful analysis called for samples to be taken from the blood of more tundra peregrines than had been documented in the whole of North America since the Pilgrims landed. To get those samples, during the early eighties Riddle headed for the Gulf. After months on the barrier islands he'd seen only a few arctic peregrines, though, and had learned that the traditional bow-net trap — which, after days of hiding in a nearby blind, might allow a falconer to snare two or three migrant peregrines — wasn't going to produce enough falcons to let him even begin his research.

Or enough for the Cancer Center to send him back to the islands for another season. He had to find more falcons, but the only part of the Texas coast Ken hadn't covered was the segment, shown as a wide lagoon on every map, that everyone assumed was water. It was not, usually, and on its vast tidal flats Riddle discovered the falcon mother lode.

It was an astonishing find. Hundreds of arctic peregrines — perhaps four fifths of the continent's entire population — had been passing down the Texas coast every autumn since the Pleistocene. But almost no one had seen them because only a few individuals came out to the beachfront. Instead, their predatory rendezvous was the Laguna Madre's tidally exposed bed, a 1,000-square-mile plain of quicksand and saline mud that hardly anyone had penetrated.

Anyone but a falconer with an ATV.

As much as peregrines meant to Riddle, however, saving them was of only ancillary interest to the University of Texas. At the time, cancer researchers viewed falcons primarily as environmental monitors: coal miners' canaries whose position at the top of the food chain made their blood and body fat a chemical sump. Their tissues, the thinking went, were where the toxic buck stopped, and from the environmentally active chemicals present in their blood and body fat a scan of the same foreign compounds that prevailed in their surroundings might be read with more accuracy than instruments could yet detect.

Peregrines also moved around, which made them a unique sort of monitor, because other carcinogen-concentrating species, like oysters and honeybees, lead sedentary lives, and the tundra falcons' biannual passage across most of each vertical hemisphere made their bodies nature's most widely ranging biotic scan. Using falcon blood platelets as a seasonal standard, it was theorized, the Cancer Center might be able to produce an intercontinental survey of the organochlorides dieldrin, mirex, heptachlor-epoxide, endrin, oxychlordane, and toxaphene, as well as of PCBs and the peregrine's original nemesis, DDT.

But to derive useful science from the falcons' chemical readout, researchers would need to know where in their travels the birds might have picked up each substance. This was the ostensible reason for Janis Chase's radio-tracked departure vectors, although it was never clear to me why her Army employers were interested enough to fund the research. Over in the Cancer Center camp, meanwhile, despite Riddle's trappers' having searched from the Texas beaches to Greenland, peregrines' northern migration routes were still largely unmapped.

South of the Rio Grande, the falcons' routes were even less known, which I found to be as magnetic a puzzle as did Riddle, with whom I had long before shared other interests for, like me, Ken had kept a boyhood collection of snakes and lizards, and it was after a late-winter lecture of mine on Texas herpetology that, for the first time in years, I'd heard his Willie Nelson drawl.

"Pretty decent job," he'd growled, waving a copy of my just-published book on reptiles, blue eyes crinkling in that tough-guy face. I said, "Thanks, and how about me coming falcon-trapping?"

5.

Renegades

Despite my invitation from Riddle it was clear that no matter how far Vose and I managed to follow a peregrine, how much we learned of its unknown ways, or how much we might have liked to join the raptor researchers, neither of us was ever going to be admitted to their elite fraternity. You had to be a master falconer, an academic ornithologist, or part of the Army brass to join what George had started calling the Peregrine Club, and as we pulled out of the Del Rio on the way to our telemetry rendezvous with my dark peregrine, George said he wasn't about to take on that kind of servitude.

I was glad to hear that, because anything he and I did with these protected birds was sure to run headlong into the norms of both the Peregrine Club and the Army, and it occurred to me that I ought to offer Vose some reassurance that I was really going to hang in there with him.

So on the drive over to Cameron I told

him how, when I was the only non-society kid at a fancy prep school, my teenage heroes had always been offbeat interlopers. Guys like George Plimpton — who played quarterback for the Detroit Lions, boxed with Archie Moore, pitched to Willie Mays, and performed as a trapeze artist with Clyde Beatty's circus — just by convincing those in charge that he could.

As a boy, that had been my goal too, and, years later, during a stint teaching literature, I'd happened to be sitting in the University of Texas Press Office when a visiting publisher's representative said he was looking for a book on Texas snakes. It was my chance to do a little of what Plimpton had done, and in ten minutes I managed to sell the university press on my writing a definitive new volume on the state's herpetology. It eventually became the 600-page *Snakes of North America*, despite my being a complete novice in the area and so shunned by the herpetological establishment that I'd had to rely on the advice of a couple of thirteen-year-old reptile buffs, who took up with me only because I had a driver's license and an off-road pickup to haul them out to good snake-hunting spots.

Vose thought about all that, then sprung

his own surprise. His life had followed a similar trajectory. After World War II George had gone back to Texas, where he joined the faculty of Texas State College for Women as a lab technician. Then, with neither Ph.D. nor medical degree, he began teaching calcium metabolism and bone histology. Just by convincing those in charge that he could.

But academia was too dull for a former combat flier, and soon George was teaching endocrinology in the morning and running Vose Aviation, a Cessna dealership and flight school, after 4 p.m.

"Had as many as twenty junior instructors when times were good. All giving lessons."

But too much unpaid flight-time and a few student crack-ups folded the little school, and George lit out for new territory. It was Texas's Big Bend country, and out there, by selling small plots along a bulldozed dirt landing strip to pilots who wanted a place to set down rough-field craft in the region's mountain terrain, Vose acquired 2,000 acres of Chihuahuan Desert.

But life on Taurus Mesa was kind of quiet, he said, so he'd taken this Army job flying telemetry on peregrines. But Janis only wanted the falcons' short departure vectors, and now that the Army was finished with even that minimal radio-tracking, George figured all he had to look forward to was the flight home.

"Least until you came along."

The morning of the dark falcon, what I found particularly likeable about Vose was the way he took to the air. It was the way other people drive. Hop in and go. Before I'd snapped my seat belt George had the Skyhawk off the ground, bucking into turbulence kicked up by an advancing cold front. Except for its lightness, Cessna '469 — the last three digits of the plane's call number, which George, like all pilots, used as its name — felt like a classic car. It even smelled like the fifties.

Not long after we'd taken off, George affectionately patted the old plaid seats. "Everything about this airplane's ancient," he smiled. "Including the pilot."

At that, 1,000 feet in the air, my door popped open.

"Wind'll pin it in," Vose shrugged.

The wind didn't do much of a job of keeping the door closed, though, and as I gazed down through its wide-angled opening I realized there is nothing like seeing the earth far below your shoes to remind you of what flying really is. So I reached out and gave the door a good slam.

The jolt popped open George's side.

He didn't seem to notice the new blast of wind, but after a while Vose leaned over.

"Photography charters, we used to just take the doors off. My old Super Cub never had any to begin with."

Even with both doors finally closed, the Skyhawk's cabin felt insubstantial: small, tinny, and hung — like the underslung passenger bay of a dirigible — by fat aluminum spars from the overhead slab of its wing. Surrounded by panels of scratched Plexiglas, everything about our little rectangular enclosure was loose and flexibly joined, like a big plastic strawberry box.

But below the wing George and I had an unobstructed panorama of the ground, which you don't get in low-wing aircraft like Pipers, whose view is upward, toward the sky. I mentioned that, and seesawing his yoke back and forth through the norther's gusts, Vose said you eventually get used to everything about a Skyhawk. Then he admitted that when he was first learning, and the wind would suddenly put his plane way up on one wing, since he had no idea where his point of no return was, it was like being in a very tippy canoe.

As we climbed high enough to receive telemetry signals, it became clear how cramped our quarters were. With every correction Vose sent a nudge into my left arm because even at the dash our cabin was less than 3 feet wide; at the seats its shoulder room was slightly narrower. But we were here in search of my dark peregrine, and I had a job to do. The Army's radio scanner, swaddled in a leather case engraved with a giant-antlered buck that suggested the rear motif of a recreational vehicle, sat wedged between the seats. Flipping it on, I pulled on my headset, fearing that once my falcon had finished preening she might have left the island. For a while I heard only raspberries of static. Then as we leveled off, 6,000 feet above the laguna, I got a strong pulse.

I clenched both fists. That beep was her: my girl was still out on Padre.

Like us, she must have run headlong into the norther, because her signal was so loud she couldn't have flown more than 10 miles from the sandspit where I'd last seen her. From where Vose and I now sat, hanging on the wind, it was easy to understand why she had stopped. Above Deer Island every squall buffeted '469 like a roadside butterfly, and to the north the indigo wall of the cold front's main air mass promised worse to come. Little by little, my elation vanished. The serene skies of my first day's radio-tracking had given me a false sense of security, and it suddenly occurred to me how on even my best days I'd never been a confident flier.

"She can't fly in this any better'n we can," I shouted over to Vose, who raised his eyebrows at the notion we were having trouble. "I think she's down for the day."

George nodded — bird behavior was my department — and heeled us over for the downwind leg back to Cameron.

During the night the rain let up, and at 5 a.m. Brownsville radar showed the storm had moved down the Mexican coast. Behind it lay 2,000 square miles of cool, stationary air. I made a face at George. That chilly upper layer would trap the laguna's evaporation as ground fog, and fog meant instrument-only flight conditions: newfangled standards, I learned over breakfast, that Vose refused to recognize. It was necessary to ignore them, too, because poor visibility would keep the dark falcon from hunting, and with neither a full crop nor a northerly wind to pin her on the coast she might migrate.

It wouldn't be light enough to take off for another hour, so George and I sipped coffee and tried to see the end of the runway.

"Same soup as on the way down here," he observed cheerfully, cranking his Cessna's imaginary yoke back and forth. "Four-hundred-foot visibility." It occurred to me that I was going to have to come to grips with the disparity between Vose's philosophy of flying and my own — which were, more or less, delight on his part and semi-terror on mine. But Skyhawk '469 was my only entree to the aerial world that a week ago had seemed so magical, and it was too late to back out now.

I mentioned some of this to George, who was of the opinion that your attitude to flying mainly had to do with how you got started. The first aircraft he ever saw was a yellow biplane that buzzed his second-grade classroom in Machias, Maine, one morning in 1929. Despite his teacher, George hung on to the windowsill and counted as the pilot did twenty-seven loops in a row. After that, it took five years of Depression-era nickels and pennies to get the $5.50 fare he split with another kid — the two of them being light enough to qualify as a single grown-up — to go up from the hayfield landing strip where barn-storming flyers came through selling rides. Wedged into the front seat of an Eagle Rock double-winger, George knew he'd found his life's pursuit.

He didn't get up again, though, until high school, when work as a hospital aide let him afford an actual lesson in an Aeronca Chief. The instructor picked him up on a horse racetrack outside town and immediately took them into a loop.

I cringed.

"No, that's how they all were back then. Tried to scare the pants off you, show you how thrilling it was. Now I take folks up and show them how safe and sensible aviation is."

An hour later we lifted blindly into the fog. Its vapor pressed like wet gray wool against our windshield, obscuring the murky world we were rushing through at 100 miles an hour, but somewhere below was the dark falcon, who by now would have left her roost and been on the wing. Being airborne should have made her signal clear, yet again she had grown as spirit-like as before her capture. Above 1,000 feet, where I'd have line-of-sight transmitter reception from all the barrier islands, the clouds were even denser, so we decided to stay low enough to be able to see, even a little. That meant we would have to search more territory, and as he broadened our listening arcs, George continued his story.

Whenever he could afford it he'd take the train ($4 for the four-hour trip) to Bangor for more flying lessons. But the war in Europe had set off a panic for coastal security, shutting down every non-military airfield within 200 miles of the Atlantic, so after high school graduation Vose headed inland. At Bellefonte, Pennsylvania, there was a flight school that offered the promise of a career in aviation and a dollar-an-hour interim job as a cutter in a nearby limestone pit. By the time of Pearl Harbor, Vose was an Army Air Corps instructor.

As soon as it was clear the dark falcon wasn't out on any of the islands we headed back toward the mainland. Because the trappers seldom saw peregrines long after sunset, they believed most of Padre's migrants spent the night over on the mainland. Maybe the dark falcon was still on her roost, so I told George to head for the Laguna Atascosa Wildlife Refuge, a semitropical forest of Tamaulipan thorn woodland so dense that within it early cowmen burned out clearings for pasture, trusting the interlaced perimeter of catclaw, prickly pear, and paloverde to fence in their long-horns. Now, 500 feet below us, those sandy meadows held whitetail deer. Startled by the sudden roar of our engine, they stood frozen, heads raised in alarm, as we were perceived and vanished in the same instant. There were collared peccaries, too, pig-like creatures Texans call javelinas, and once in a while a dark-humped nilgai — a looming Dr. Doolittle antelope-confection gone wild courtesy of the Valley ranchers' penchant for exotic game. Gazing down through the mist was so interesting I almost forgot it was the only direction we could see; horizontally, there was only a dun-colored opacity.

At the edge of the refuge it was clear that we had taken the wrong heading. Beyond were crop fields, where no peregrine would spend much time, I told Vose, pointing across the laguna. Probably we missed her on the island.

George nodded. "Less to run into out there, anyway."

Losing a little altitude as we came back over the water, we squeezed into a narrow band of clear air between the cloud bank and the bay, dotted here and there with spill islands — football field–size deposits left from dredging the Intracoastal Waterway. I knew that not all the offshore falcons went over to the refuge, because in the headlight of my tugboat ATV I had seen peregrines perched on driftwood washed up on the spill islands — safe out there from both the coyotes of the barrier beach and from the great horned owls that everywhere in their range seize sleeping peregrines in their outsize feathery feet. Owls are uncomfortable over open water, though, and on the laguna's mudspits a falcon could doze securely right out in the open.

Except for a flock of pelicans the spill islands were empty, but somewhere over their small archipelago my headset clicked with a single faint beep. Janis had said that if a transmitter got wet it could partially short the battery, which might be why we had lost most of the dark peregrine's signal. Yet as a few more dim beeps came in, their wavering tone told me she was in the air, and nosing the Skyhawk right and left like a quail-scenting pointer, Vose and I listened our way up the coast. Soon the dark falcon's signal strengthened, and I realized that her transmitter hadn't gone bad; it was just a long way off, with beeps that were loudest from the northeast.

That meant out over the Gulf. I pointed seaward, but George shook his head. Flying blind in the fog was one thing, but venturing away from land with '469's rickety ignition would have been too risky even for him, so all we could do was angle north along the breakers, periodically banking oceanward to see if the dark falcon was still out there.

Forty miles up the shore her transmitter pulses had grown loud enough to hear without earphones and, maps spread, I had an answer. Our peregrine wasn't over the water at all; she was all the way past Corpus Christi, where the Texas coast bends eastward. To have gotten that far ahead of us she must have gone aloft long before dawn, something none of Chase's other migrants had done.

An hour later, as we, too, angled east above the upper coast, the fog had come down a few more notches, putting our wheels almost on the surf. But ahead of us the dark falcon hadn't faltered: she was aiming for Louisiana. Following nineteenth-century biologist Constantine Gloger's rule — that the pigmentation of animals from warmer and more humid habitats tends to be darker than that of races of the same species living in drier and colder climates — then pale-breasted peregrine populations inhabit dry climates; ventrally dark ones, humid regions. All of which meant that my dark peregrine's coloring indicated she was probably a member of the southern *anatum* race and therefore could be heading for anywhere in eastern North America.

And so, it seemed, were we. Except that, only 8 or 10 miles behind her now, we couldn't see a thing. My stomach knotted: soon we would have to leave the open air of the coast and, like our falcon, fly inland over terrain filled with signal towers, buildings, and other aircraft. None of that would hinder her, because peregrines' crucible eyes are somehow able to penetrate fog. Falcons don't usually hunt in such conditions, but the trappers had told me that on migration their drive for home sends them rocketing through even the heaviest weather.

That seemed to be Vose's inclination, too, for as the mist ahead of us finally met the water, he didn't waver.

"Might as well have a little altitude," he offered as the Cessna's nose came up. "Never saw a snowstorm I couldn't fly through in twenty minutes."

Forty minutes later we could just make out the ends of our wings. Our altimeter showed 900 feet, and from our flight chart it was evident that we were about to enter the air-traffic corridor between Corpus and Houston.

Cameron Flight Control had warned us that in similar weather, just weeks before, a pair of military pilots had rammed their jets head-on just off the nearby naval air station runway, and as I opened my mouth to remind George, a dim red light flashed past. Then there were other red beacons and blinking white ones, on both sides, above and below us, all of them marking the 1,000-foot-tall radio towers that flank the region's petrochemical refineries.

I stabbed at the chart, motioning "up," and in seconds Vose, who was jerking the plane right and left like a video game ship, had pulled us clear of the towers.

As we climbed, I was thankful that for once George looked shaken enough to have grasped the seriousness of our situation, and after a while he folded his hands and looked down. There wasn't anything visible through the windscreen anyway.

"We aren't going to be able to do this, Alan. I can't see to get down for fuel."

Ahead, the dark falcon signaled steadily,

set at last on the 12-degree heading that would carry her along the route I'd envisioned since we'd turned northeast. Those firm wings that just hours ago had struggled against my hands might, over the next two weeks, carry her up the Mississippi Valley. Over the Great Lakes, maybe deep into Canada. It was a journey Vose and I would not share, and in silence we watched the dashboard's floating-bubble compass bob back around to the familiar heading that would take us home.

6.

Sport of Kings

Coming in to Cameron, the Skyhawk didn't break out of the clouds until we were almost over the runway, but I'd left the receiver on, since there was still one radio out on a falcon. It wasn't carried by one of this spring's tundra birds; it was worn by a partially crippled first-year peregrine that had just been released by Riddle's second-in-command, a veterinarian named Bill Satterfield.

Vose looked over when he heard the signal.

I tilted the readout so he could see her number. George recognized the frequency: he'd been listening to her beep for more than a week, using its easily located pulse to check our antennas' reception, because, since she had been on her own, Satterfield's peregrine hadn't moved from a small area near the end of the south island.

It was almost certain that meant trouble,

for a stationary peregrine is one in difficulty, and this little falcon's well-being meant a lot to me. I'd been at the beach cabin where she was brought in back in October and like many of the newly flighted youngsters that arrived on Padre, she was thoroughly beaten up. But Crazy Legs, as we called her for her twisted, previously broken left leg, was worse. Somehow she had made it down from the Arctic on wings gapped by the stubs of broken flight feathers, likely crumpled by tackling too-large prey, and in desperation she had come in to one of our pigeons.

Satterfield, who was both a vet and a student of falcon ecology, wasn't optimistic. Turning her skin-and-bones body carefully in his hands, Bill shook his head. In her condition, Crazy Legs was barely able to fly, much less capture prey with her dysfunctional leg, and releasing her would surely mean her doom.

"Even if she lives, I can't see her successfully defending a nest ledge. And if she doesn't reproduce, statistically she's a questionable member of the population." There was nothing much that could be done with crippled peregrines, so with a dismissive shrug Satterfield slipped off her body stocking and for the moment set her on one of our carpet-covered perches.

Squeezed between stacks of supply crates and big olive duffels, Crazy Legs trembled beneath her hood. Our size had overwhelmed her, left her a thing of mere beauty, bereft of ferocity — a fragile fabrication of keratin and hollow bone vulnerable to the crushing weight of every piece of equipment that lay piled around her.

She couldn't even look at us. In the Cyclops' cave of our beach cabin, the sight of her lumbering captors would have provoked constant, frenzied flight, and in the darkness of her hood she could only bend to pick at the bits of pigeon breast we slipped beneath her toes. Otherwise, Crazy Legs did not move. Satterfield was able to stand that for almost two full days. Then the scientist in him capitulated, and to soothe away her fear Bill knocked off his trapping.

His work began at night. An hour after the rest of us had turned in, I heard him begin to cluck the low, infant-feeding call peregrines use as baby talk with their mates. Mingling his calls with the pulse of distant surf, Bill plied his courtship during the long hours of darkness, pouring out his counterfeit peregrine voice, like Cyrano, from the shadows. Even with her hood removed, his small Roxanne could not see him well enough to panic, and by the third night she had started to relax, cocking her head expectantly as he began his serenade. Deep in my down bag, I wasn't sure when I first heard her reply.

By the end of Bill's second week Crazy Legs was won. Unhooded in the cabin's near-darkness, with eyes once more agleam she sat confidently on Satterfield's glove, pulling chunks of meat from its leathered fingers and spitting defiance at everyone but her chevalier.

When the trappers left Padre early in November, C.L., name shortened to initials and prognosis still indefinite, went home with Bill. I didn't see them again until February, when Satterfield called from his Bastrop County farm to say it was time for C.L. — contributing member of the breeding population or not — to begin preparation for a launch back into the peregrine mainstream during spring migration.

After so long in captivity she'd never get near a prey bird, though, until she'd regained her predatory keenness. That was what we had to give her.

Pulling on a khaki shooter's vest stuffed

with lure, leash, and transmitter, Satterfield stepped a now more muscular C.L. onto his familiar welder's glove and headed down into the back pasture. With her hood lifted, C.L. blinked, shook out her feathers, and preened a bit. Bill released his end of the deerhide tethers that dangled from her ankles and waited. C.L. twisted her head back and forth, scanning the field for meadowlarks. Not seeing any, she minutely inspected her talons. It was a routine the two of them had followed for a month, and C.L. was in no hurry.

Finally, to the jangle of her ankle bells she lifted off and circled overhead, waiting for Satterfield to swing the lure, a meat-carrying leather approximation of a flying bird she'd have to swoop at a dozen times to get her dinner. It was the only thing that bound them, yet unlike the wild-caught passage bird she had been four months before, C.L. had no interest in the larger world. A flock of blackbirds swept by above her, but she hardly noticed.

"Still no confidence," Satterfield muttered, readying the lure as C.L. flailed around on lumpy wings.

Then she changed. Spooked by her circling shadow, one of Bill's barnyard pigeons fluttered up from the grass, and suddenly, bad wings and all, C.L. was bound to the chase. It wasn't much of a chase, as the pigeon streaked into the shelter of Bill's hayloft, but it was enough.

"I'll be damned, she got a flight," Satterfield said. "If I can fix those primary feathers she might make it after all."

The surgery took nearly an hour.

C.L. never knew she was on the way under as Bill lowered a plastic cone over her head and turned on a whiff of anesthetic. X-rays had shown that the break in her left leg had healed at an angle, leaving that set of claws marginal for hunting, but there was nothing that could be done about it. All Satterfield could do was help her fly. Peregrines are so valuable that parts of dead individuals are preserved for recycling, and though he'd temporarily patched C.L.'s wings so she could go aloft for exercise, Bill hadn't had what he needed for a permanent repair. Now the set of pre-owned wings he'd been waiting for had arrived from the Peregrine Fund.

Clipping away C.L.'s damaged flight feathers — five from the left wing, four from the right — Satterfield left a half inch of quill projecting from the skin. The stubs were hollow, and he carefully reamed each one to the outside diameter of the shaved-down shafts of the deceased peregrine's quills. Then he twisted the new flight feathers up into C.L.'s empty sockets and aligned them with what remained of her wing surface.

"Imped-in primaries have to be a perfect aerodynamic match," he said, squinting along the overlapping layers and making minute adjustments as he wicked a drop of Super Glue into each feather juncture.

Next came curved suture needles and fine stitching twine. "I got this thread from a guy who manufactures shoes in Massachusetts," he said, sewing together, for good measure, each of the nine slip joints. "It's lightweight, weather-resistant stuff that absolutely will not rot."

As Bill moved to the falcon's crumpled tail, the stainless-steel door behind us cracked open, and Riddle, unable to stay away from anything concerning one of his group's peregrines, eased into the operating room.

"We found a bird on the beach last spring with only three tail feathers," he told me. "He was small, even for a tiercel, and he was struggling to stay up, so I looked all over for substitute retrices. Cormorant, or even gull, would have been preferable, but the best we could come up with was a bunch of pigeon feathers. That tail looked pretty funny when I sewed it in, but he had a long way to go and really needed the lift. We caught him again a few days later and he seemed to be flying all right."

Crazy Legs' new tail wasn't going to be an exact match, either, since it consisted of the wider, grayer tail feathers Satterfield had saved from his own fifteen-year-old falconry-trained peregrine's last molt. But with a bit of trimming the new retrices fit well enough, and after stitching them in, Bill doffed C.L.'s oxygen cap and his own surgical gear and carried her into his office.

For a long time she lay cradled in his arms, breathing deeply.

"Look at this silly gal," he said. "Just snoozing. She doesn't know she's off the anesthetic."

He flipped C.L. to her feet, and in an instant she was transformed. Kindled once more into a smoldering carnivore, she pierced the room with her blazing eyes, hissed, and with flailing wings leapt away from Satterfield's wrist.

The next day I got a phone call.

"C.L. made a kill!" Bill yelled. "The dogs flushed a scaup off our pond while she was up, and she just rolled over and took that duck so clean you wouldn't believe it."

It was only five weeks until spring trapping began on Padre, and with Satterfield I went down a day early. A gusty south wind greeted our newly fettled ATVs as Bill and I pulled away from the beach house, riding slowly because he was steering with just one hand; on his wrist, proudly hooded as a falcon in a Restoration woodcut, stood C.L.

Swaying confidently with the Honda's every jounce, she gripped her leather perch and leaned into the turns, wings folded serenely on her back from long practice. It was their last ride together. At noon, a CH-47 Chinook from the Army base at Harlingen was to meet us at Mansfield Channel to ferry the first trappers over to a camp on the inaccessible northern island. Bill would be going with them, after he released C.L.

Our chopper was already on the ground when we reached the channel, and its pilots, glamorously bored in their zippered flight suits and black aviator glasses, were tired of lounging in the shade. The second they saw C.L. their impatience vanished. It was time for pictures: each officer holding her, hooded and unhooded, a close-up of the transmitter stitched to the underside of her central tail feathers, and a group pose on automatic timer.

Bill cast around for more pictures but couldn't put it off any longer, and slowly we rolled out onto the flats. Three miles southeast he stopped, removed C.L.'s anklets, and slipped off her hood. As usual, she shook herself out and swiveled around, eyeing the vast horizon.

"You're free, Gal," Satterfield said softly. C.L. looked down and nibbled the place on her yellow shanks where her tethers had been. Then she dropped from Bill's wrist, swooped an inch off the sand, and swept away with the wind.

It had happened too quickly and, insanely, we leapt on the ATVs and spun off after her, burning out the emotion over the sand. For more than a hell-for-leather mile we could see her flickering wings before she vanished into the mist over the dunes.

That night, over his third or fourth gin and tonic, George told me that since his bad knees kept him from going out on an ATV, he'd never seen a wild peregrine up close. He had plotted enough of their departures with Janis, and followed others —

105

with her and earlier researchers — far enough to have piqued his interest in where they went, but Vose said he still didn't understand the fascination that falconry held for every trapper.

I told him I thought it might be because human beings had always been drawn to horizons they couldn't reach on their own — horizons the flight of falcons gave them a taste of — although falconry certainly didn't start that way. Using birds of prey to hunt began, probably in Central Asia, more than 3,000 years ago as a purely practical way to capture small game and waterfowl. Though falconry was unknown in Greece and Rome — in Cicero's *De finibus*, the recognized occupation of bird-catcher was limited to small songbirds taken in nets or mired in sticky lime — an Assyrian bas relief from the ruins of Khorsabad, built in the eighth century before Christ, shows a hunter with his hawk.

Coursing with raptors was well developed by at least 600 B.C., when the earliest written accounts of it as a means of gathering food appeared in China. But by the first millennium A.D., the functional side of hawking had been eclipsed by its spectacle. Like racehorses, peregrines and gyrfalcons became the passion of Indian rajahs, Japanese shoguns, and Arab sultans. In 986 Marco Polo reported that the Kublai Khan derived

the highest amusement from hunting with gyrfalcons and hawks. At Changanor the Khan has a great Palace surrounded by a fine plain where are found cranes in great numbers. There he causes millet and other grains to be sown in order that the birds may not want.

On his annual hunt at Changanor, Polo reported, the Khan and his entourage were accompanied by 10,000 falconers, able to send hundreds of hawks simultaneously into the air — a sporting extravaganza, other travelers noted, soon copied by royalty throughout the Middle East. According to the Fuertes, who also wrote of epic historical hunts, the Sultan of Bayazid kept 7,000 trained hawks, monopolizing his realm's supply because, like beautiful women, birds of prey were so desirable that one not belonging to the royal retinue was an affront to imperial prerogative.

It was from those great Mid-Eastern lofts that returning Crusaders — men so taken with the selfless bravery of falcons that during later Crusades they spent much of the time between battles engaged in hawking — brought the first trained falcons to Europe. Yet falconry did not catch on.

Like the sixteenth- and seventeenth-century mariners, who ferried home lions, gazelles, apes — and one phenomenally popular giraffe — long before their captives' needs were known, the crusaders with their hunting hawks were little more than a curiosity. Despite their periodic infusion of trained falcons, transferring to another culture a discipline as exacting as that of falconry was difficult, and in most of Europe hawking was still a rudimentary pursuit as late as the tenth century.

Then the sport, and soon the hawking way of life — the commoner's means of putting game on the table, the aristocrat's badge of status — became a cultural sensation, seizing the Western imagination so powerfully that until the advent of firearms late in the seventeenth century, hunting with birds of prey was rivaled only by religion in the intensity of the passions it ignited.

Without question, the single most important force fueling this phenomenon was the ardent participation of two of Europe's

most prominent citizens. Frederick I —
better known as Barbarossa, the first Holy
Roman Emperor from the house of
Hohenstaufen — was both an avid falconer
and a wildlife enthusiast, implementing
Europe's earliest conservation laws: stat-
utes that prescribed harsh penalties for
harming a hawk.

But Barbarossa's grandson was the real
fanatic. In Germany, the second Emperor
Frederick built an entire castle just for his
falcons, to whom he was so devoted he
once lost an important battle against the
Tartars when he failed to show up for
combat because he was out hawking. Un-
deterred by his loss, Frederick kept up his
connection to the Oriental caliphs, whose
kingdoms lay just to the east of his own, by
importing large numbers of trained
peregrines and gyrfalcons, as well as eagles, from as far away as almost-mythical
Cathay.

And whenever he could, Frederick also
imported his birds' handlers, because
hawking in thirteenth-century Europe was
innocent of both sophisticated Asiatic
training methods and the immense body of
art and literature that had accumulated
during the thousand years that, in the East,
captive birds of prey and their human han-

dlers had worked together.

Bringing this wealth of art and learning into the Holy Roman Empire was Frederick II's life's work. During the thirty years of his reign he authorized the transcription of whole libraries of falconry texts, footnoted with lessons learned in his own training lofts, then molded everything he had discovered into *De arte venandi cum avibus* (*The Art of Hunting with Birds*).

Due largely to Frederick's patronage, after the publication of this influential treatise, the most desirable birds of prey came to be valued so highly that Philip the Bold, Regent of France, was able to ransom his son back from the Saracens with twelve white gyrfalcons. But *De arte venandi cum avibus* was far more than a blue-book of raptorial skills.

Completed only after Frederick's death in 1250 by his son, Prince Manfred, *De arte venandi* dealt not only with the complexities of practical falconry, but, almost irrespective of its subject matter, was one of the first important examples of postclassical European literature, eventually rising to greater prominence as a work of literary scholarship than as a manual of sport.

The reason lay in the way the great book

defined the language of learned discourse. Frederick's faculty of translators rendered scores of Asiatic languages and dialects into classical Latin — making falconry's Eastern terminology available to Europeans — which was complicated because there were few equivalent terms in either Latin or any other European language. Thus the Emperor's translators had to invent entire scholarly dictionaries to interface their myriad disparate lexicons. Yet with decades to work and unlimited funding, Frederick's scholars eventually succeeded so well that, substantiated by their patron's standing as both Roman Emperor and scion of the venerable Hohenstaufen lineage, their work was eventually accepted throughout Europe as both the prevailing international linguistic reference and the standard of literary excellence.

Among those who could read Latin. Meanwhile, several social notches below this learned aristocracy, among the Continent's newly monied gentry the hawking life was beginning to gain favor. Despite the fact that almost none of its participants could decipher the arcane vocabulary of *De arte venandi*, the book was widely recognized as the apogee of sophistication, which opened an opportunity for translator/interpreters, none of whom was more popular than British abbess Juliana Barnes.

Over the centuries Barnes — or Berners, as her name sometimes appears — has become a legendary figure. Perhaps the daughter of Sir James Berners, she has been portrayed as being of gentle birth and to have been an actual abbess, possibly of Sopwell Nunnery in Hertfordshire. She was also almost certain to have been both the first female author of a printed work in English, and, until J. K. Rowling, by far the most successful.

The abbess's 1486 treatise, *The Boke of Saint Albans*, was, except for the Holy Scriptures, the most widely read work of its time. During the 130 years following its publication it was reprinted no less than twenty-two times, for Barnes was the first to address in print the newly powerful, yet mostly untitled Tudor social class that was to dominate English society for the next four and a half centuries. And to those who, because of their common origins lacked both the spoken French and Continental graces of William the Conqueror's Norman followers, *The Boke of Saint Albans* served as a defining social text, with Abbess Barnes as the Tudors' Emily Post.

A century later, Barnes's word on etiquette, social mores, and especially proper vocabulary (including the language of falconry) was taken even more religiously by Elizabethan society. In *Every Man in His Humour,* urbane Ben Jonson, always dubious of the new rural gentry, wrote:

Stephen: Uncle . . . can you tell me, an he have e'er a *Boke* of science of hawking and hunting; I would fain borrow it.

Knowell: Why, I hope you will not a hawking now, will you?

Stephen: No, wusse; but . . . I have bought me a hawk, and a hood, and bells . . . I lack nothing but a book to keep it by.

Knowell: O, most ridiculous!

Stephen: Nay, look you now: Now . . . a man [that] have not skill in the hawking languages nowadays, I'll not give a rush for him; they are more studied than the Greek, or the Latin!

According to the Right Hon. D. H. Madden, author of *The Diary of Master William Silence,* for such an ambitious young man, the task of scrutinizing Barnes's *Boke* "was no trifling one. There was a separate word for every conceivable act done by, or to, each beast . . . with an endless array of appropriate verbs, nouns, and adjectives, the misapplication of any one of which stamped the offender as no gentleman."

Shakespeare, too, acquired much of his "gentle" speech from the huntsmen and falconers with whom he spent his youth. Then, with all the authority of *The Boke* itself, he made use of this linguistic badge of sophistication by employing in his plays and poems almost 250 hawking words and phrases, many of which had been first translated by Juliana Barnes from the Latin in Frederick's *De arte venandi cum avibus*.

Although *The Boke* also dealt with a host of social issues unrelated to falconry, its most famous passage is Abbess Barnes's proposal of a fictional legal system assigning every bird of prey to one of fifteen social classes. Pretending her strictures constituted actual law, the abbess used hunting hawks as ciphers to codify the disparate strata of society. Those who flaunted her codes by keeping sporting birds deemed to be beyond their social station were soon to be set right, she wrote hopefully, by having their hands cut off.

According to Barnes's made-up legislation, squires should fly only the less-spirited lanner falcon in order not to encroach on the higher hunting standard of their knights, for whom it was appropriate to train the fiery but less reliable desert-living saker falcon. Up another social level, dukes, earls, and barons could hunt with peregrines, but only with the smaller male tiercels, because princes alone could fly the big female peregrine.

Noblewomen were expected to take pigeons with pretty, graceful little merlins — one of which Mary, Queen of Scots was allowed to fly throughout her captivity — while churchmen above the rank of priest were allotted the practical but unspectacular meat-hunting goshawk. Wealthy commoners nevertheless often owned and flew peregrines of both sexes, a practice Barnes approved of only if the family held an ancestral coat of arms, which in practice usually was a historical fabrication acquired at great expense.

Would-be gentry unable to afford such revisionist lines of descent were to fly only the little hobby falcon, the abbess went on — though, despite *The Boke*'s popularity no one steeped in hawking paid much attention to it, and finally, mostly as a joke, to peasants and clerks Barnes assigned only the mouse- and sparrow-catching European kestrel.

The largest and rarest of the falcons, the pearly arctic gyrs, were reserved for kings, while only emperors such as Frederick, maintained the abbess, were to ride afield with wolf-killing eagles on their wrists. Except in literature, however, eagles were mostly for show, because when nobles went hunting, like everyone else they took peregrines. An inveterate falconer, Henry VIII became so fond of a favorite tiercel that he almost suffocated when, in pursuit of it, he tried to vault a canal and had his pole snap, plunging him headfirst into the mud.

That such a thing would be readily understandable to Henry's peers indicates how thoroughly, by Elizabethan times, sporting with birds of prey had permeated both English and Continental society. Hawks went everywhere, sitting hooded on perches set up in dining halls and sleeping chambers, appearing on the wrists of sporting couples as they exchanged vows of marriage, and accompanying so many clerics to the altar that bishops finally prohibited lesser ecclesiastics from bringing them into the vestry.

Yet among all this avian bestiary, only peregrines and gyrfalcons became symbols of high romance.

Their regal looks were part of the reason, but what also elevated peregrines above other sporting livestock was their bravery. Properly conditioned, a 2-pound female would swing boldly into battle with an 8- or 10-pound goose or heron, heedlessly flinging its delicate body against its huge opponent in the air. Only a peregrine or gyr would face those odds.

And only a peregrine or gyr could succeed: if they were skilled enough, and had their timing just right, the best falcons could flash out a killer hind-talon at the end of a 150-m.p.h. dive to slice the head or wing off any aerial prey. If it missed or was knocked away, a prime hunting peregrine's fiery spirit would send it towering back up to attack again and again, even if that meant dying in its final assault. Eventually this otherworldly courage was linked to a newly minted aesthetic of aerial balletics — a wing-borne dance of pitch-and-roll, throw-up and power-stoop — and for centuries the mortal mix of art and combat incarnate in the flight of falcons obsessed the civilized world.

7.

Chemical Warfare

"All that history doesn't tell me a thing about why these Army fellas keep coming down here," said Vose, having whittled his Gilbey's down to its bottom third. "Much less why they're being so secretive."

He got up to see me out the motel door, remembering to point out that whatever the Army was up to, the two of us weren't going to have one of these high-powered little hawks to follow if I didn't get a radio on another one pretty quick.

So, long before dawn I drove over the causeway back to the island. Even at sunup no one would still be at the cabin, because the trappers who bunked there had to be 30 miles up the island by first light, but I stopped in anyway to check the logsheets. Daily accounts of falcons tried for, missed, and captured, they were the trappers' contribution to the Cancer Center's data bank, and we left them in a communal stack for one another to read. Since the crew from

North Padre came in only every other week, it was how everyone stayed in touch.

There wasn't a lot of news. With storms still flooding the lower flats, nobody had been catching falcons, so Riddle's young assistant, John Hoolihan, had taken a Zodiac across the channel to join the group on the north island, where there was higher ground. Then, before I went out, I called Cameron's flight office. It would mean a lot to Satterfield if I could retrap C.L., whatever difficulty she was in, and I wanted to see if George had found her in the same spot on his morning flight. He was still up, though, with a passenger.

It was my girlfriend, Jennifer Miller, who had come down to visit. With me gone out to the flats, though, George was showing her the lay of the laguna, as well as how — by homing in on C.L. — we picked up telemetry signals, so, in search of C.L. I set off up Padre, looking for other peregrines along the way.

I didn't get far. Last week's inblown tide had left an abrasive paste of salt that, mixed with wet sand, had worked its way into my Honda's drive-chain, gradually corroding its already rusty links into a loop as rigid as a welded mailbox support. It took most of the rest of the day to hike back to the cabin for a new #520 chain, and by the time I got it installed several trappers' mud-clogged ATVs had pulled up beside a newly arrived rental sedan. Inside the cabin, all the beer cans were in the trash and Hoolihan was policing a week's worth of used dishes.

"Ward's here," he said.

I ducked at the news; after so long eating combat rations and taking quasi-military orders I almost believed I was part of the Army contingent, and Lieutenant General Prescott Ward was its commanding officer. When I'd first come out to the island, Riddle — who seemed to buy the Army's line about being involved for the public-relations value of helping endangered species — had consigned me to Hoolihan, who walked a fine administrative line between the sometimes disparate objectives of the university's cancer researchers, Riddle's single-minded falcon trappers, and the assortment of military types who showed up from time to time. As a result, John seemed to have evolved a policy of saying as little as possible about anyone — and since nobody ever questioned my background I assumed tight-lippedness was a tradition in the program and bunked for days with guys I knew absolutely nothing about.

Just now, John said, Ward was out fishing. I could see him at the water's edge, a compact man with short brown hair, flinging a big silver lure. It was hard to figure. Ward was a real falcon biologist, but I had heard that he was also part of the Army's Chemical Warfare Division. There were plenty of martial-engineering devotees in the biological sciences, and I knew many of them were drawn to the speed and power of raptors. One of their projects, the trappers had told me, surfaced at a conference where its architect had explained how he'd raised a nestling peregrine, feeding it on a scale model of a high-rise office tower while repeatedly showing it a video of the flight path leading to the actual building. When the newly fledged falcon was released nearby, the researcher was proud to announce, it flew straight to its designated target.

Everyone from chief Army veterinarian Bob Whitney to Satterfield and Chase had spoken well of Ward, and I'd seen his name as co-author on research papers. The Army's portion of Riddle's program might really be on the falcons' behalf, I decided, so I headed down the beach and waited for Ward's lure to come in.

The general looked over at me, then out

at the horizon. Without a word he flung his heavy casting spoon again, 60 yards past the surf into deep water. As his lure wriggled slowly back though the waves I fidgeted, then stepped forward and thrust out my greeting hand.

Ward stopped reeling and looked at me.

I told him that I'd done some nature writing. Maybe I could help with his project's public relations.

"Help?" he barked. "Every magazine and newspaper on the East Coast is trying to find out about this program. Prying for angles. One of our falcons got in a scrap with a local peregrine back in Virginia and the Audubon people were up in arms."

He rolled his eyes.

"Then there's the animal rights jerks."

The thought upset him.

"Morley Safer and *60 Minutes* want to come down here — may already *be* here — and film a segment."

He aimed his fishing rod at my chest.

"Now, get this: *nobody* talks to them."

"But isn't that the point? If you don't want media attention on this — on conservation — well, why's the Army radio-tracking peregrines?"

Ward lowered his fishing pole.

"Which unit are you with?"

I told him a little of the truth: I wasn't with any unit, I was here to help with the banding.

"Who are you, anyway?"

It was clearly a bad idea to keep on with the part about being a nature writer. A journalist, like Morley Safer. So I asked Ward if he'd caught anything with that big chrome spoon.

Sooner or later Hoolihan would have to spill the beans, though, so I went on.

Vose and I were on their side, I told the general. Paying all our own expenses, helping the project's peregrine research. No cost to either the Cancer Center or the Army, except for the radios.

"Radios?" Ward gasped. "You have transmitters on our birds?" He grabbed his reel and started to crank.

"Lost the last tagged one up by Corpus," I said unhappily, watching the line singing onto his reel. "Now we're just monitoring a sick rehab bird."

"No you're not," Ward choked. "Not anymore!"

Right in front of me, the pact that George and I had forged — my only chance to enter the transcendent world of a migrating peregrine — was falling apart. The general was so angry I thought he might try to have me arrested on the spot, and sure enough, he took a couple of strides up the slope toward the cabin, then turned.

"You, sir, may be looking at a federal offense. Now, stay exactly — precisely — where you are standing."

I wasn't quite sure what Ward's relationship with Riddle was, but as I sprinted for my truck I knew I had to get Vose and the plane out of Dodge, and in a hurry.

It took nearly an hour to reach the airfield, and I had time to think. Without Vose there could be no project. But he'd piloted for this Aberdeen Proving Ground bunch during two previous trapping seasons and had only just met me. Much as he'd seemed enthralled with our plan, I was sure George would change his mind when he learned what had happened. Ward, after all, was a formidable adversary, this was his pet project, and we were, at the very least long-overdue, permissionless borrowers of his equipment.

The other problem was the transmitters. Even if C.L. was still alive — our signal could be beeping from her dead body — she was apparently never going to leave Mansfield Channel. And now that I was off the beach due to Ward, we had no way to get a radio on another falcon. Somewhere up on the north island a trapper named Mike Yates was carrying transmitter #.759, but it had been more than a week since I'd given it to him, and since he hadn't been particularly enthusiastic about the non-routine procedure of attaching it, I had pretty much written him off. Besides, the minute Yates met another trapper who had gotten Ward's instructions, that radio was headed straight for the general's brief-case. That left only #.465, tucked in my pack.

By the time I got to Laguna Vista, George and Jennifer were back from their reconnaissance, settled by the flight-office coffee machine. As I opened the door, Jen flew out of her chair.

It was great to see her, except that, deep in each other's arms, we couldn't see much. But as I pulled back to kiss her, she got a look and wrinkled her nose.

A week on the flats had left me covered with layers of salt spray, mud, and pigeon blood.

"Sweetie! You're a piggy mess!"

She tried to pull away, but I wouldn't let go.

"Essence of Alan."

She rolled her eyes and, leaning back, dug a handkerchief out of her cutoffs to mop at my face. I wagged my head back and forth like a little kid, but I didn't let go.

Jennifer made a scary face and, knowing I'd hang on, pulled back really hard. The same height, and as small-boned and freckled as me, she'd always felt like a female other self, and this was some of our favorite stuff; in a second she gave up and moved closer.

Over her head, marooned by so much physicality the guy working Laguna Vista's flight control was looking for an escape. But Jen and I had the door blocked.

"They OK?" he asked George.

"There's people here, Babe," Jennifer whispered. I kissed her and moved away.

"You two, uh, get reacquainted now," said George. "I gotta refuel the plane."

I waved him back: there was important stuff he didn't know.

"Son-of-a-bitch general flew in last night. Threw me out of the Army. Not that I was ever *in* the Army. But out of the program. Our deal's dead."

Vose shook his head, genuinely dismayed at having riled anyone. But he didn't falter. He drew himself another cup and said he'd been in tougher situations, lots of times.

"I'll call Janis. Straighten this out."

Chase wasn't at her Maryland condominium, so the three of us caucused. Everyone in the peregrine survey knew that George and I were based here at Cameron. It wouldn't take much for Ward's guys to show up and put us out of business.

Vose stroked his mustache.

"Just for the time being," he began, then stopped for a deep slug. "I think we ought to clear the hell out."

In minutes there was 100 feet of air between our wings and the runway. Safe for the moment, but with no real destination, the three of us headed across the laguna. We'd cleared the island and come to Padre's ocean beach when Vose looked over for where to go next. With no real reason I pointed north, along the surf, and George banked left. I'd turned around to check on Jen just as she noticed that on the backseat next to her lay the last transmitter. She picked up the package and looked inside.

"I don't suppose it's possible to drop something out of one of these little planes?" she said. "I know the FAA probably doesn't. . . ."

Against the press of wind, Vose was

prying open the hinged Plexiglas panel next to his elbow.

"Throw out anything we want," he yelled over his shoulder.

There was plenty of streamer left on George's roll of surveyor ribbon, so Jen and I started to work. By the time I'd written my note she had packed the transmitter in shock-absorbing bubble wrap. We were taping Day-Glo streamers to its package when '469 arrived over North Padre, where I was sure that none of the guys had been in contact with Ward because I was the one who was supposed to have fixed their two-way radio.

"No reason we can't locate one of those dune buggies," George observed, dropping to 200 feet. "Got both tanks full, only so much island to cover."

We didn't need a quarter of it. I was still adjusting to the video-game effect of our low level speed-scan when a set of ATV tracks flashed beneath our wheels. Ten minutes later, we roared over an olive tent, the trapper's base. Wrong direction. Fifteen miles the opposite way the red dot of a rolling ATV flared, trailing a pale wake of tire-flung sand, into sight. This is what we look like to a peregrine, I thought, as George swept down the machine's incised path, and as it grew larger we could see that the Honda's cargo box was white and twice as large as the other ATVs'. That meant it was Hoolihan: ol' trap-'til-you-drop, never-come-in-while-you've-got-a-swallow-of-water John Hoolihan.

Tossing our tiny package into the vastness of the flats would be chancy, but if anyone could spot its fall, it would be John. Seeing us, he trailed out a strand of toilet paper to indicate the wind direction, then stopped dead with his back to the breeze. As we shot by, skimming the flats like a falcon, George brought the Cessna around, down to 40 feet off the deck. Into the wind with full flaps we dropped to 65 m.p.h., the stall horn blaring its gut-tightening squall.

"Don't mind that thing," Vose instructed, wedging open his window. At 200 yards, out of the haze John reappeared, crouched behind his handlebar. George yelled, "Now!" and leaning forward from the backseat, Jen pushed out our trip's last hope, a fluorescent lump the size of a softball.

It was beautiful. Through the rear windscreen I saw the Skyhawk's propwash tumble our package upward against the beige sky, then bloom out its steamers into a Day-Glo octopus that spiraled slowly toward the sand. As we turned, climbing, it collapsed orange-tentacled on the beach, 100 feet in front of John's Honda.

Port Mansfield was the nearest landing strip, and having stopped there on the flights with Janis, we knew it had a phone. This time George reached his boss. He pointed out we might be able to add a lot to what she'd learned about falcon migration, and that we would take good care of the Army's equipment. It didn't work: Ward had lowered the boom.

"Don't understand that fella." Vose sighed. "But I guess we haven't any choice."

It seemed incredible that we were through, yet as George pulled us off the runway we were headed back toward Cameron. A Quintana Oil Co. Twin was coming in, and as we circled to get above it I put the receiver on C.L.'s frequency. Radio still working, but I wondered if she was really alive. Then, just for the hell of it, I hit the scan switch. In the receiver's little crystal window the numbers spun through their gamut of possible combinations, then stopped on #.465. It was the transmitter we'd just dropped to Hoolihan.

Maybe John was fooling with the off-on

magnet. As .465's beeps faded, we turned, and they came back on with the wavering pulse of a falcon in flight.

I yelled at Vose and Jen, and they whooped at me. John had done it. An hour after our drop we had a transmitter bird in the air.

I took over the Cessna's controls while George unfolded his charts.

"Robstown," he said after a bit, pinning the airport under a big wrinkled finger. "One way or another, I'll stall 'em a couple more days."

A hundred miles up the coast, Robstown's Coastal Bend Aviation was our safest bet. Ward's guys would be checking the landing strips adjacent to South Padre, but to the north lay nothing but empty grazing land belonging to the King Ranch. If we based ourselves at the northern end of that wilderness, we'd probably be out of the Army's search area, yet would still have a chance to intercept .465 if she migrated up the coast toward us. It was a somewhat plausible strategy, though I was sure that Ward wasn't going to give up on his equipment even if he didn't find us right away.

Maybe we could buy our own system. The manufacturer's name was right on the receiver: Telonics, Inc., Mesa, Arizona. From Robstown, I called them. They had a similar model, for $4,500, but company policy was to deal only with universities, research facilities, or the military. Didn't want their equipment being used, the salesman said, by suspicious spouses tracking each other's cars.

That made it a case for Bob Binder. Longtime friend, Austin mayoral candidate, and attorney, Bob had worn a lot of hats on my behalf. This time I was asking him not only to advance me more than $4,000 but to pose as the Chihuahuan Desert Research Institute. The real C.D.R.I. was run by a friend of George's, Dr. Dennie Miller, but its principal advantage to us was the fact that the organization, beset with financial problems, seldom had anyone available to answer its phone. Bob said he was willing to try, and with a single persuasive call to Telonics he shuttled the Chihuahuan Desert Research Institute from the campus of Sul Ross State University to his law office in Austin. A certified check was sent, and as soon as its system could be programmed with our frequencies, one brand-new model 3610, TR-2 receiver/TS-1 scanner was to go out via UPS.

Meanwhile, we found that Hoolihan's peregrine was a different sort of traveler than the dark falcon. Perhaps only recently arrived from the south, she didn't seem ready for another journey. At night, she hid away in Laguna Atascosa's thorn thickets, crossing the bay at dawn to feed on the offshore islands. Afternoons, she dozed on the flats. Before long, George, Jennifer, and I learned to look for her signal along the beach after her morning kill, then fly back to the mainland to gas up the Skyhawk. Late in the afternoon, we'd follow her home to roost.

Afterward the three of us would head north — often with Jennifer learning the controls — back to Robstown's Royall Court, where we'd check in just in time for George's cocktail hour. Across the street, someone on the city's recreation committee had designed a circuit-training track of chinning bars and sit-up boards that circled the municipal park, so leaving Vose to his charts, Jen and I usually went over.

Because it was wonderful. Strolling arm in arm through the gauzy spring twilight, the town's well-fed Hispanic community blithely ignored the fitness stops as they circled the track in the traditional evening paseo. Merging into their flow, Jen and I took our place among middle-aged couples and pretty girls, all of us circling past observant pods of slicked-down field hands and white-shirted clerks. Every day that we managed to spend with .465 seemed perfect, and giddy with success, I said that if only she would take us farther up the coast, we might be able to follow her forever.

Jen gave me a look.

"Alan. Think. Next birthday you'll be what? Forty-six, forty-seven. That's *middle-aged*. Are we really Bonnie and Clyde?" She tugged at my arm. "And what if this bird doesn't migrate? Never goes anywhere? I've got a job to go back to."

I said I knew she was a family case worker, and there was a limit to what others could do to cover for her at work. Still walking, I gave her a long, jiggly kiss on the cheek, but she didn't buy that either.

"No. Now, listen. How long can we actually elude this man? This *general?*"

I listened to Jen take a deep, frustrated breath. It was hard for her to make points to me, and I hoped she was about to give up. Instead, she took a different tack.

"Hon, can we afford this? George has no money. . . ."

As if to tell me a secret, she leaned up to my ear.

"Plus, you know how you are with Authorities. . . ."

Authorities actually meant something to Jennifer, which infuriated me. But before we got into that again, a plump Hispanic couple eased around us, compensating for our lack of a common language with exaggerated smiles and nods. Jen just sighed.

"Look at these people. They're contented."

"They're bored," I snapped. "A week here and you'd be nuts."

She looked at me, with something like yearning, I guessed, though there was a hard edge to her voice.

"No," she said. "They're happy. Like we ought to be."

8.

On the Sea Wind

From the curb, I watched Jennifer's back cross the road to the Royall, then did another lap. When I got back, the light was out in our room so I went next door, where George's entrance was open to the night air. With aerial maps spread on the floor and all over the bed, Vose sat at the table flanked by a quart of gin and a jumbo bottle of tonic water. With his 28-ounce Big Gulp cup, he waved me in.

I got a bathroom glass and poured a dollop.

"Glad you're here," he announced. "We been needin' to talk."

I swallowed the gin. I'd been afraid of something like this, but no matter what Vose wanted, I couldn't lose him and Skyhawk '469.

"Mister," he said, "I been thinking about heading home. Taking off here already lost me my Army job."

He deliberated a minute.

"Now, that Jennifer, she's got a real touch at the controls. But I have my expenses. . . ."

"Expenses?" I barked. "Who was it going on about intestinal fortitude? About hanging in there?"

Vose was calm. "I'm just not sure any of this makes sense."

He paused.

"Even to you."

I poured myself another g & t and waited for him to continue.

"Now we've finally got this bird on the radio. . . . Well, you can't even see it from the plane."

Inside, knots had formed. Without Vose and the Cessna I had no dream. No way to discover the peregrines' aerial migration route, that path through the sky that nobody — not one researcher in the world — had plotted to its final destination. Without George I would never share a falcon's life, up in the clouds, completely apart from the world of man.

I checked the cord to the Army receiver, made sure the battery was getting its nightly charge, and went into the bathroom to polish off the rest of my drink. Then I shoved the maps out of the way and sprawled on the bed.

Vose hadn't spoken, but, stern as Vermont granite, his long Yankee features remained fixed on his new heading. "Matter of fact," he said suddenly, "back in Machias — that's Maine, where I grew up — we all shot hawks. They even paid you: bird of prey bounties sent me to many a movie matinee."

"Difference in generations. Yours loved chickens."

"Oh, we valued wildness, too," Vose shot back. "Just not when it was eatin' our livestock. And we had so much of it — so much open land — maybe we took it some for granted."

I nodded. "When I was twelve," I said, "my dad shot a hawk. Knocked it out of a tree at a hundred yards with an iron-sight Remington .22."

Bright and undiminished, the young redtail had twisted onto its back as it fell, hooked feet cocked against the scarlet stain spreading across its breast. My dad, my brother Rob, and I scrambled out of our canoe. Then I halted. From the shoreline gravel the hawk's burning eyes had riveted my own. I could not look away, and for a long time I did not move. Then the liquid sulfur of its gaze dimmed and my father,

138

thinking the hawk was dead, stooped to pick it up. But again its eyes flamed to life, and with a hiss the redtail scythed three glistening talons through both sides of my dad's hand.

He screamed and ripped away the claws, the hawk's flailing wings showering us all with blood. Yet, propped once more on its tail, it had glared up, not just unafraid and ready to fight, but seething with maniacal rage.

Nothing had ever stood up to my father: not our hunting dogs, not even Blackie, his Angus bull. Certainly not Rob or me. Incandescent in its vehemence, that gun-shot redtail became magical. It showed you didn't have to give in, even to your old man. Even if you were about to die.

Then its golden orbs faded, but as they closed I swore — in an inner voice stronger, bolder than my own — to keep them. Keep those eyes, always flashing, in my sight.

Vose set down his Big Gulp.

"OK, Mister," he sighed. "I get your point. These little things findin' their way all over the world — I guess that oughta mean something to an old flier too."

Mornings, it was a long haul down to the

islands, but from high above the rug of gray coastal cumulus, Jen, George, and I grew used to finding .465's signal along the eastern shore of the laguna. That let us set the plane down to wait, usually at an unmanned strip whose runway was right on the water. Next to the airstrip, along the muddy verges of the bay, sanderlings were common. Within days they, like the peregrines that hunted them, would resume their long flight back to nesting grounds in the Arctic. For the time being, though, they were engaged in their own predation: minuscule black legs and feet a blur, each bird shuttled from snail- to wormhole in individual pursuit of hidden invertebrates. Then an old Beech Bonanza rumbled in, and as the sanderlings flushed into the air they were transformed. In an instant they ceased being ten separate birds and became a single shimmering cloud flicking back and forth over the water. Sweeping in unison through every instantaneous dive and turn, each sanderling maintained an exact 6 inches from those that surrounded it. We all shook our heads.

"How can they do that?" Jen said. "They must have a leader, but what tells the others where he's going to go?"

"Formation flying," observed Vose. "Worst mistake you can make is to change your mind — and these little things do it every half second."

Being tiny probably helped, as did experience. In the bunched peloton packs of bicycle racing, I'd been in trains of closely packed competitors that at times could seem to be a single creature, snaking right and left through downhill switchbacks and up sudden climbs, the whole pack standing together out of the saddle.

At best, we could equal the sanderlings' speed and maintain a proportionately close spacing. But we had pavement to follow. To reach the sanderlings' order of cooperative magnitude you had to be more than trained; you had to *be* exactly alike.

Descartes saw this, and deemed animals to be machines: closely replicated, almost identically responding mechanized beings without self-awareness. He was probably partly right, at least about fish. As a teen working with a film crew in the Caribbean, I was entranced by schools of silverside minnows whose members would invariably part just enough to keep my outthrust hand from touching any one of them. With no pushing, no broken ranks, each rearward tier of fish had darted backward exactly the distance their fellows in front needed to move to avoid my fingers. Somehow, all those near-identical neural systems gave each of their owners perfectly equal flight stimulus — and nothing else.

Later philosopher/naturalists were re-pelled by the notion of higher animals as non-sentient mechanical cogs. Henry Beston theorized that shorebirds were linked by waves of mystical energy. That is yet to be established, but life in the wild forces animals to be efficient, and none channel their energy more effectively than small migrant waders, for creatures as light as sandpipers are so buffeted by every puff of air that they need one another to fly well. Drawn together in spring and fall, their tiny pelotons share the task of pressing into the wind, where each indi-vidual relies on the help of its companions to conquer the hemispheric distances they travel. Breaking apart for foraging, where individual initiative works best, our Mansfield flock had instantly re-formed for the next micro-leg of its journey north.

But to pull that off a hundred times a day means no extra baggage. Any mental circuitry the least bit more complex than that of your companions would set you apart. It would give you more choices, let you change your mind in different ways. That would make you the odd man out — and in the sanderlings' world, oddities don't survive. Peregrines see to that.

Among the trappers, unusually plumaged pigeons were prized because the asymmetrical flash of one differently feathered wing excites falcons more than almost anything. True reactionaries, peregrines are always on the lookout for the slightest aberrance in a member of the wading flocks they regularly flush into the air. Often it seems the falcons put their prey to flight just to see if one of the shorebirds is flying unevenly, for any non-conformity is likely to mean the bearer lacks fitness or coordination and is therefore an easier catch.

The result is sameness.

Evolutionary incentives toward homogeneity have been going on so long and so relentlessly that almost every wild creature has been genetically pummeled into a what-works-best uniformity. Falcons themselves are no exception. As with other hawks, their sameness involves not only their looks but their temperaments. To hide, to cower, to appease when the odds are too great seems to humans to be an obvious choice. "A living dog is better than a dead lion," according to Ecclesiastes, but that is only because the option to behave as a metaphorical dog or lion is there for us.

Those options aren't part of a hawk's repertoire. Carrying the potential to make either choice would be excess baggage — additional neural circuitry that, even if a raptor could evolve some way to fit the software in and manage its extra weight aloft, would hurt a bird of prey's chances for survival. Too many alternatives would slow its swift predatory reflex, while an injured hawk on the ground is so likely to die anyway that its species would not benefit from having additional capacity for evaluating different contingencies.

Without even the possibility of a cowardly response, then, a bird of prey makes no conscious choice to resist an overwhelming threat, exhibits no grace-under-pressure force of will. Those are mankind's concerns.

In falcons are other things. Agendas of flight and killing, allegiances to icy cliffside places, and the pull of the old circadian timings — of equinox and solstice, cycles of light and darkness that in concert with a thousand other subtle imperatives still beyond our knowing send them, twice every year, across half the globe. Yet, irrelevant as it was to their own lives, those bold medieval peregrines demonstrated that a swift, delicate creature could completely disregard ignominious fear, thereby making the choice of human bravery a vivid, living possibility for every soul that ever revered them.

As usual, after our long midday stop, by evening .465 had made her way back to the mainland, and in the morning, full of expectation, we arrived at her roost by 9 a.m. She was gone. No signal at her favored rest sites out on the island where we'd always found her. She couldn't have migrated north on the same heading as the dark falcon because every day we followed that route down the coast, scanning for her signal all the way. There was no way she could have slipped by us.

"Not unless she left way before sunup," Vose said. "None of the ones Janis and I followed ever did, but that might have been how your dark falcon got so many miles on us."

Paralleling the islands, we flew on south. Twenty miles below her feeding grounds, it was clear .465 hadn't just moved down the beach on a prey chase. George pulled us around, and with slim hopes, we headed back. Our greatest fear was that, as often happened with Padre's falcons, she had been trapped again, which meant her transmitter would have been repossessed by Ward's troops. For two hours Jen, George, and I hardly spoke. Corpus was behind us, Houston ahead, and despite a 360-degree listening turn every few miles, we heard nothing.

There was nowhere else to go. Stupidity, ineptitude, bad luck: we had lost another one. Hoolihan's falcon, our last transmitter bird. With a final sweep, we headed back to Robstown. There, at Coastal Bend Aviation, a message was waiting.

All week George had held center stage at the airstrip's flight office. That hadn't seemed wise to me, but fifty years of aeronautical exploits had generated too many good stories for either Vose or his audience to pass up, and inevitably word had leaked. The message was to call Riddle at the university.

When I hung up I felt terrible. I'd been afraid I might lose Ken's friendship, but what had happened was worse. Despite the potential damage to his program's funding, Ken had called only to warn us. There was a chance we might be sought by the police.

I didn't mind the cops — that risk was worth it — but harming a friend who had been generous enough to ask me along in the first place was intolerable.

That was it, then. The Army's equipment had to go back. Our falcon was long gone anyway. Ward had been so angry I didn't think simply returning his radio-tracking gear would patch things up for Ken, but I needed to try. So before Jennifer left she and I packed up the military scanner and took it over to Robstown's bus station; disassembled, the Army's long-range antennas went, too, but not before I'd taken their measurements. If our new receiver ever arrived, George and I would need new antennas, even if the only bird we had to locate was C.L.

Ken had told me .465 hadn't been retrapped, so her loss was entirely our fault. We didn't know where she had gone, though we were pretty sure it wasn't northeast up the coast like the dark falcon. But before we could follow another peregrine we needed to identify the flaw in our tactics. I'd checked the vectors of Janis's last ten telemetry departures, which had seemed to put us right on the falcons' main route north.

But it was not their only route. A couple

147

of non-vectored mavericks had peeled off early. If .465 had flown *northwest* from the brush country where the falcons roosted she would have been 80 miles inland and out of radio range by the time she passed Robstown — set on a course aimed like that of at least two predecessors, straight at the spine of the Rocky Mountains.

It was another basic miscalculation. Peregrines went where they pleased, and if we ever got on to another one, George and I saw we would have to follow it all the way to bed, then be there in the morning before it woke.

That left us with a plan and no equipment. But that night, to what Binder had designated as the Robstown campus of the Chihuahuan Desert Research Institute, UPS delivered our new scanner, and with its arrival our enthusiasm began to rekindle.

Packed next to the receiver, Vose and I found a pair of stubby, handheld little Yagi antennas designed for monitoring small mammals at close range. They were nothing like the Army's big tubular Christmas trees, but C.L.'s signal had always been easy to find, so we hose-clamped the rubber handgrips of our new Yagis onto the Skyhawk's wing struts and went up for a test.

We were hardly off the runway when I had a beep.

The plane was still too low for it to be C.L., 60 miles away on the banks of Mansfield Channel, but George heard the signal too, and right in the middle of takeoff he reached over and gave me a thump on the shoulder.

"Four six-five! She came back."

But she hadn't. In the scanner's window was locked a different number: .759. It took a minute to remember. I'd asked Mike Yates, up on the north island, to put the transmitter I'd given him on a strong adult female, and he apparently had trapped one, for this peregrine was moving up the coast so fast we figured she must already be on migration. With that in mind, Vose and I stayed right over her the rest of the day, keeping within the 15-mile reception range of our little mouse- and rabbit-telemetry antennas.

But .759 wasn't going yet. At mid-afternoon she began to work her way south, and by 6:30 she had made it all the way back to roost in the same mainland oak thickets the others had used. By then, she'd flown so far up and down the laguna that even with a refueling stop George and

I were low on gas and had to scramble for an airport.

The nearest was Falfurrias, an agricultural crossroads 70 miles inland. Neither Vose nor I had ever been there, but I liked its remoteness even before we touched down. Surrounded by miles of empty pastureland, its airstrip consisted of two short runways, a turnoff, and a concrete-block office the size of a single-car garage. A cattleman's rough-field Cessna 180 was the only plane in sight.

Falfurrias Municipal was run, single-handedly, by Dan Wilson, a cruiser-weight individual who could hardly have been further from the mainstream of civil aviation. Falfurrias was a lonely place, Dan said, and he was delighted to see us. He told us he liked falcons and had watched them on television, but that he and his wife were more into rocks, especially minerals, and gemstones in particular. It was a difficult hobby, though, since he hadn't had a day off in seven years.

George was so dismayed he offered to take Dan up for an outing in the plane right then and there. But after eyeing the moonscape of tiny craters punched into '469's wings and tail by two decades of West Texas hailstorms, Wilson ran his fingers over the antenna cables draped from the 2 x 4s that sandwiched our wing struts and allowed as how he'd probably spent enough time in the air already.

Besides his rocks and documentary TV, Dan was, as he put it, a Back-Roads Scholar. Also a sometime bird-watcher, he added, so after talking it over, George and I decided to let him in on the deal.

Falfurrias was going to be our hideout. Our Hole in the Wall. Radio #.759 was still Army property, and we knew from a message Riddle had left with Jennifer that even with his receiver back, Ward remained unmollified. Wilson said that as far as he was concerned he hoped our falcon would take up permanent residence around there, and while he might not go so far as to actually lie on our behalf, if anyone came snooping around while we were up he'd have a hard time remembering if he'd seen us.

He never needed to forget, because Falfurrias, located in the midst of dairy country and therefore home of the Fighting Guernseys, was a great little town. George and I spent nearly a month there and never heard an unkind word, never were served a bad meal, and never had a problem with the law. Out by the highway, the municipal airport sign was missing half its nailed-on letters, but it didn't matter, because the 8" x 10" shingle was stuck behind a utility pole and you couldn't read it anyway. We stayed at the Antlers Inn, where taxidermied heads were the primary motif, did our laundry at the Llave-Mat, and every morning drove out to the airstrip past the high school's statue of an enormous Fightin' Guernsey, blood in her scowling eye.

But Falfurrias Municipal was a long way from our peregrine. Counting takeoff, it took Vose and me half an hour and 4 gallons of aviation fuel to reach .759's feeding grounds, and once we'd located her, in all the surrounding 120 square miles there was nowhere we could land to spend the day monitoring her on the radio. That meant we had to remain in the air, expensively burning fuel, then fly back to Falfurrias, talk minerals with Dan for a couple of hours, and head back out to the barrier islands for another telemetry reading before dark, hoping .759 hadn't migrated in the meantime. Halfway home on the third day, I noticed a thin line of pavement just east of Highway 77.

"Private ranch strip," George observed, tilting us over for a look. "Narrowest one I ever saw: like somebody's driveway. But it's long."

Vose threaded the needle, and as '469 set down our falcon's signal was still strong on the scanner. It was hard to pay attention, though, because we had come to Shangri-La. In the midst of South Texas's sea of gray thorn brush a thick St. Augustine lawn undulated beneath ornamental palms that, next to a kelly-green lake, looked like they belonged in Florida. Beside the airstrip a hangar held a big blue Cessna Skywagon, and on a rise beyond the lake stood barns and a red-tile-roofed hacienda.

At what the veranda's leaded-glass doors informed me was Rancho Sarita, no one answered my knock, but through the panels I could see rooms of memorials: a pair of full-size red, white, and blue cowboy-boot bookends, plaque-mounted steer horns, and big framed photographs of well-dressed people on podiums. Richard Nixon was front and center.

"Que quieres, Señor?"

The question came from a trio of lean brown men drawn up behind me. They were polite but had little English, so we relied on my halting Spanish. The owners weren't home, but there was a number I could call; their name was Armstrong.

I should have known. The huge ranching operation that spread for miles on all sides belonged to Tobin and Anne Armstrong. She had served as the first female ambassador to Britain, Nixon's emissary to the Court of St. James. The Armstrongs' son had for a time been a falconer, and my brother Rob knew their daughter, Sarita.

Armstrong was surprised when I called him, but gracious and generous. We could use the strip.

The only difficulty was with our gear. If .759 ever really took off we'd never be able to follow her with Telonics's short-range Yagis, but just up the road was Alice, an oilfield town full of rig suppliers. I was sure someone there could weld up copies of the Army's stainless Christmas trees, but two afternoons of dragging from one semi-shut-down shop to another made it clear that nobody had stainless tubing — or was interested in fabricating it if they did.

Meanwhile, the weather closed down. In Falfurrias, George and I didn't turn on the morning report anymore. Even in the dark you could look outside and see the wet gray fog, and hearing the regional flight office's instrument-only restrictions just irritated Vose. Of course we were going up.

We had to, though with that 600-foot ceiling we were hanging on to .759 by a thread, and even with the benefit of being able to make short loops out of Armstrong's field we were right on the brink of disaster.

The solution came from Bill Mading, an old motorcycling buddy and a welder. His Austin shop already had more work than it could handle, but when he heard what we needed, he pushed aside a stack of orders and spread out the drawings Jennifer had brought him. The next evening, two 6-foot stainless steel antennas, taped to cardboard backing, boarded a Greyhound for Falfurrias.

At last we were set, except our falcon wouldn't migrate. It wasn't that she was sedentary. She flew all the time, dragging us back and forth through that terrible morning mist nearly as far north as Corpus, then south again across the laguna and South Padre, where we stayed too high for Ward's trappers to spot the Skyhawk, hoping that .759 wouldn't go after one of their pigeons.

Afternoons, if it was sunny, our girl defied the birding guides that declare peregrines seldom soar by fanning her wings and wide triangular tail, riding the ranchland thermals up a thousand feet to swing slow, non-hunting arcs with the turkey vultures. From Armstrong's strip, with our duplicate military antennas we could follow her course around each ascending spiral, tell when she'd left the vultures to flicker out on her own, and practically take her pulse.

It was intoxicating. Without this creature's being disturbed by us, or even aware that we existed, we were coming to know her, gaining a feel for her characteristic patterns of movement and feeding — what she would do in the rain or when the sun broke out. I was sure no one had ever lived with a wild falcon like this, thousands of miles from its nest ledge, at the midpoint of its solitary pilgrimage across the continents and, deadly morning fog or not, every day George and I grew happier in our connection to her.

Meanwhile, our own characteristic patterns were being logged. At the Antlers, Maria or Arlene would have George's eggs and my oatmeal on the burners before we got to our customary corner booth, and every morning Rosa would stifle a giggle as we lugged our duffels down to the desk to, once more, check out.

"See you this afternoon," she'd call from

beneath a forest of deer horns as we heaved our bags into Dan's loaner car. "Don't fly away too far."

Out of deer season, the Antlers was deserted except for Vose and me, but sometime after midnight there was a knock on our door. It was Binder. After daily calls to keep up with our new receiver's performance, the weather, and any additional way that his bogus Chihuahuan Desert Research Institute could continue its role in our project, Bob had decided he could no longer take the suspense of being left out.

"Brought you your mail," he announced, wedging in. "Maybe we can go up in the morning."

He was in luck, for by the last week in April the early spring fog had cleared, leaving Kenedy County awash in full, cicada-song summer. Above Armstrong's runway, after we'd located .759 and set down to wait, the humid air was so still I could hear the scritch of a column of big red ants that wound beneath my camp stool. It was the only thing there was to look at, for after her morning kill, out beyond the coastal dunes .759 dozed on one of her usual driftwood perches. Under Armstrong's ancient oaks Bob studied his legal briefs, and in the shade of the

Skyhawk's wing George flipped through the novelization of a movie for which, the summer before, he'd done the stunt flying.

After a while Vose tired of his paperback and got back to thinking about falcons. How come, he wondered, with all the organized training people put into them, hadn't hawks ever settled down? Like dogs?

It was a good point. After more than thirty centuries of seeming to work as partners with people, why were falcons still entirely undomesticated? It would have been easy enough, even in mankind's earliest days, to raise stolen fledglings on scraps of meat, and to tether them as adults. Contemporary Amerindians keep parrots, monkeys, and coatimundis tied outside their shelters. But hawks won't reproduce in that kind of captivity, which meant that no successively tamer generations could have been produced; every wild-born nestling would have been as emotionally remote as its free-flying kin.

More important, hawks won't defer to humans. Popular accounts depict the primal wolf/man bond growing out of a predatory alliance — human hunters relying on the wolves' scent and speed, the wolves on primates' spear-power. Imagined on some frosty plain, it makes a dramatic illustration, but it's unlikely to be true. Our relationship with canines, which biologists call inequiline commensalism, probably started, instead, with pups brought home from dug-out dens. Or, yet more likely, scavenger jackals simply began hanging around campsite dumps.

Either way, as submissive/dominant pack animals, little canids are genetically predisposed to appease their group's leaders, and the supplication of an extended paw or moist nose-nudge works even better on innately empathetic humans. Rewarded by scraps, the tamest and most submissive camp followers would have stayed, bred, and eventually physically ceased to be quite wolves at all, retaining into adulthood the smaller brains, jaws, and shorter legs of adolescents. In exchange for security, dogs probably chose to trade away their wildness by domesticating themselves.

The psychology of that wasn't lost on the ancients. An appeaser, a coward, was a dog; a far more fragile hunting hawk, on the other hand, was an icon of respect. For one reason: its soul was inaccessible.

You could capture a bird of prey but, even in a world of casual cruelties, you couldn't break its will. It either died fiercely, without terror, or it flew away like a spirit. At most, and only with infinite patience, you could make a bargain with it — an arm's length, help-each-other-kill falconer's agreement. For no matter how accustomed a hawk becomes to its handler, in its heart it never sees its human as anything more than a way to satisfy its need for food.

The remoteness of falcons, I went on, their essential untouchableness, was one of the reasons I thought they meant so much to Riddle. And maybe to me.

George nodded, but he was still having trouble with the idea of how so many people he knew pretty well could be so completely caught up in falconry.

"Just seems sorta contrived," he said. "Even those power dives — never saw one, so I could be wrong — but they've got to be brief. In one place."

Brief and in one place wasn't Vose's style.

"You know, they're only stunts. Like some aerobatic jockey, compared to a genuine aviator. Lindbergh. Or Earhart — now, there was a gal."

The idea gave Vose momentum.

"I think the main thing this little hawk — falcon — is going to do is migrate. Real, long-distance flying."

"This peregrine's a female, right?" Binder broke in. "And we know she's on some extreme journey?"

He beamed.

"Then why not call her Amelia? Amelia Long-Distance Earhart."

9.

Amelia

Vose, Bob, and I had just opened our tuna-can lunches when we saw the limousine coming down the hill toward the airstrip. Like every other vehicle on the ranch, it was beige, and it had no doors. I'd never imagined a limo without doors, but despite the deer-rifle scabbards attached to its black vinyl seats, the doors' absence made it easy to exit, and a tall man in pressed gabardines stepped smoothly out, bearing a pleasant, vaguely distracted air. It was the first time I'd met Tobin Armstrong.

"Still listening to your bird," he observed, gesturing at the Skyhawk's open cabin, where a rubble of empty saltine wrappers and plastic soup containers covered the floor. Subtly, I slid one of my rough-weather garbage bags over the taped-up crack that split our rear windscreen.

Patron of the surrounding 80,000 acres of Tamaulipan oak-savannah, Armstrong hadn't seen '469 before, and as his gaze swept absently over the pocked surface of the Skyhawk's stained wings, it was hard to tell what was going through his mind.

George never noticed. From the far side of the plane he clumped enthusiastically around to Armstrong, hand outstretched while he was still 10 feet away. The rancher looked at me.

"This your pilot?"

I nodded, realizing that nothing about us fit Armstrong's expectations. Not our battered Cessna, and certainly not Vose, who was far too upbeat, too self-assured, to be flying a ramshackle little expedition like this. His accent was pure Down East.

"George P. Vose," he announced, pumping Armstrong's arm, "transplanted Mainer. Guess you know we've been tracking a radio-equipped peregrine falcon."

Reaching into the front seat, George waved our rattiest set of earphones at the rancher's head.

"Have a listen!"

Armstrong raised a hand in refusal, smiled a brief acknowledgment at Binder, who was still waiting to be introduced, and wheeled away in a spray of gravel.

"Nice fella," said Vose. "Little strange."

By noon even the cicadas had hushed.

Halfway down the runway a pair of blue-birds flipped in and out of their fence-post nest box, wild turkey pullets bobbed through the nettles at the edge of the thornbrush, and 20 miles away, like the tick of a grandfather clock grown familiar as a heartbeat, every 1.8 seconds Amelia's transmitter clicked its small sonic pulse through our speaker. Prone on the laid-back passenger seat I dozed in the heat, a teenage kid back in Florida, watching my hunting hawk spiral up against the sky.

Then something was wrong. George and Bob had gone back to their books, our makeshift windsock hung as lifelessly as it had all morning, and the faint hum of static continued to buzz on the receiver.

But that's all there was. My heart starting to thump, I pushed the Cessna's tail down to raise its nose wheel, pivoting the lightweight craft on its tires. As its antennas completed half a compass arc, at 310 degrees — inland — a single, barely audible blip made its way through the static. Then there was silence.

Vose had looked up to watch me turn the plane.

"She's going."

North? He gestured, and I nodded,

though neither of us had the slightest idea where Amelia was about to take us.

We hadn't planned to level off until 5,000 feet, but I got a good signal as soon as we turned northwest, and after a couple of right/left sweeps to pick up her heading, at 90 m.p.h. we took up the chase. As the Skyhawk, still climbing, broke through the upper layer of cumulus, George and I could see the highest of the vultures Amelia had been circling with just beginning to augur up into the clear.

For the last week, she had spiraled endlessly in their thermals, buoying up her pound-and-a-half torso on wings that from tip to tip spanned more than 40 inches. I had wondered what she was doing up there, for her slow daily circuits signaled a sea change in the focused hunting hawk that, every dawn for nearly a month, had raked across the barrier flats in search of prey. Maybe this was some sign of vacillation, or even conflict: a struggle between her growing need to follow the lengthening daylight north and her inclination to keep to the safe roosts here, close to the easy hunting of the flats.

As a mature bird Amelia would have faced this choice before — maybe many times — and would know of the days to come, crossing hostile terrain, barren of prey, and of the nights, haunted by the great killer owls that waited everywhere along her route. Perhaps that was what had held her here so long. Or maybe it was another memory — a recollection of the late-spring storms that, spilling down from higher latitudes, still waited to roll across her path.

At least that was what I imagined. Or maybe, non-sentient after all, she remembered nothing, was just restless, or even — as Descartes would have maintained — was exhibiting nothing more than increased vernal flight activity.

But then why those towering gyres? Why seek the vultures' thermals every morning, riding up thousands of feet beyond the clouds where no prey bird ever ventured? And why go higher every day? All Amelia could find up there were vantage points — ever loftier aerial sites from which her laser eyes could pick out the upthrust highlands that, 100 miles inland, would guide her north.

Within her there had to burn a force — some image more ephemeral than a blueprint of the genes, something more dreamlike than memory — but some awareness, some close-held picture of a rocky cliff-side place not far from where she herself had hatched. Gradually the pull of that distant prospect must have swum into focus, pulling her up toward the heights she would need to vault the continent. And somehow, this afternoon, far above Armstrong's salt-grass pastures, the three of us had managed to be with her at the moment when that primal vision at last exploded across her consciousness, spinning her away like a bright spark flung from the circling vultures.

Wisps of chocolate cloud, the last of their flock swept beneath our wings but, 5 miles ahead, Amelia was already changing — not just choosing her course of migration, but starting to remake herself. As the minutes passed it was like seeing a butterfly slough its gray-shroud chrysalis, for with every mile Amelia was creating a new winged being. The mostly idle, afternoon-dozing little hawk George and I thought we knew so well was leaving us, becoming an unfamiliar 30 ounces of singular purpose. Stroking deeply and steadily, she swept past the shoreline, and, focused now on lands far beyond our familiar coastal haunts, like a departing angel pressed on toward those distant uplands.

This was what I'd glimpsed on my first flight with Janis — the primordial flame of which Vose and I had only a vague idea when we chose it as our beacon. Now, caught in its power, the two of us exchanged a look. Locked on this heading, Amelia had committed herself to the western fork of the tundra peregrines' long highway home. If she lived, with or without us she was going to Alaska.

For a while we flew in silence. Then the realization of what we had accomplished hit Vose, and out of nowhere he let out a long whoop of elation, shoving the wheel in and out to porpoise roller-coaster swoops across the sky.

That bounced Bob off the backseat he shared, gingerly, with the spiky prongs of our two extra antennas. But that wasn't his main worry. He had planned on only a morning hop and his car and all his clothes were back at the Antlers.

"I'm not sure," he began, "that it's safe to leave my Mercedes. . . ."

"Gotta travel light, Boy," George hollered over his shoulder, wagging the Sky-hawk's wings as we rose and fell. "We're hooked up! It's a Nantucket sleigh ride!"

It was more than an hour until Vose and

I calmed down enough to realize that, after swinging a series of 5-mile figure Ss to keep from overflying Amelia, we were low on fuel. The land had greened, and ahead, across irrigated fields Amelia's signal was still strong, so I'd gotten out the charts to find our fuel stop. But as I was about to pull off my headset I heard her signal falter. It was only a waver, yet for the first time since we'd left the coast, Amelia had angled away from her original course. Ordinarily that would mean she was casting back and forth in search of a place to perch, but this time she was gaining altitude, and as she began a climbing circle we caught up and, half a mile above, slid past.

Sprawled like a pale, scaly fungus, across our forward horizon lay San Antonio, its skyscrapers outlined in an ash-brown cloud.

"See that?" George pointed. "It's pollution. People move out to my part of the country, come back for a visit, and tell me they forgot how cruddy it was."

The almost-hidden city roiled with traffic we couldn't yet see, but its exhaust fumes had already permeated the Cessna, and I tried to imagine what, from 300 feet above the ranchland mesquites, that unnatural fog, resonant with the throb of five o'clock freeways, looked like to Amelia. All I could tell from her signal was that she seemed reluctant to enter it.

Comparatively speaking, this wasn't the bad stuff. For decades Texas has produced about a third of the nation's toxic waste. Nearly half of that comes from petrochemical plants located between Houston and Lake Charles, Louisiana — an area natives refer to, without irony, as the Golden Triangle. Saturated with unrefinable chemicals, along the Triangle's ubiquitous waterways live mollusks and annelids that thousands of migrating wading birds pluck from the mud. Petroleum residues weaken some of the migrants, making them easy kills for northbound peregrines, and the chemicals concentrated by the shorebirds' metabolisms add their bit to the stew of organochlorines many of the falcons are already carrying back from the Tropics.

Going inland so early had taken Amelia away from the upper coast's chemical sloughs, and I was glad to see her pull back from even the San Antonio smog; Vose and I didn't want to fly through it either.

Pausing for Amelia to decide what to do, George turned right, setting us in on a smooth asphalt strip near the village of Poteet. Besides Lonesome Dan's pair of isolated runways, Poteet was my first rural airport, though places like it had been George's transitory homes for decades. He'd told me how at these fields everybody flies, pumps gas, and, certified or not, at one time or another teaches aviation and works on planes. At Poteet we could hear Amelia's signal clearly, even from the ground, so for a few minutes we could afford to look around.

Forty miles out from the city, it was sunny and quiet on the runway, with a mocker warbling in the pyracantha next to the flight office. The people were just as nice. Edna and E. T. Page, two mechanics, a flight instructor, and both teenage Page daughters trooped out to look at antenna-sprigged '469. Buying fuel, George was in his element.

"You're going to Alaska?" Mr. Page asked, after Vose's story. "In that?"

George glowed with pride. Only a long-distance light-plane specialist, it was understood, would even consider such a trip. Vose and I both knew our talk about flying to the Arctic was mostly fantasy, for even if we could stick with Amelia as far as the Rockies, the Skyhawk's 11,000-foot ceiling would keep us from following her through the high passes that the U.S. Fish and

Wildlife Service had told us peregrines invariably sought.

Yet even if we managed only to hang on for part of that high-altitude ride, to have already shared Amelia's life every day for more than a month, to have been with her at the moment of her decision, and to still be accompanying her well away from the coast seemed miraculous. It must have looked that way to the Pages, too, because after everyone had sproinged our antenna tips and taken a turn listening to Amelia on the headsets, E.T. solemnly shook our hands.

"Wish we were going with you," he said. "Every one of us."

When Amelia's signal dimmed, seven hands waved us back into the air, and before we'd cleared the end of the runway I could tell Amelia had made another choice. Only a few hundred feet above the ground, keeping the traffic to her right, she paralleled Loop 410 as it circled the southern part of the city. Three thousand feet above, Vose and I exchanged grins. To the north the metroplex stretched away for miles; this was the best possible route past its urban sprawl, and our girl had figured it out.

Two hours later, far beyond San Antonio, Amelia bent back north. As 310 degrees bobbed back around on the Skyhawk's floating compass, right over the spot she would have passed on her original heading, Amelia swerved to the left, picking up her old course. For a migrating bird, often believed to be at the mercy of winds and following only generalized, instinctual directions derived from starshine, that sort of precise geographical orientation was a surprise, but in the months to come George and I were to see peregrines do it again and again.

Ahead lay different terrain; a dark-green uplift filled the western skyline. It was the Edwards Plateau, the southern edge of the limestone hills Amelia could have seen from high above the coast. From the plateau's rim, oak-capped ridges fingered down into the surrounding fields, suggesting an inlet-fissured shore — which, at least five times since these rocks' formation as Cretaceous seabed, those bordering limestone hills had actually been.

When we started radio-tracking, Vose had told me it was nearly impossible to see falcons from the plane, but he had watched golden eagles from the air, and, once, a peregrine. Now, with less to occupy my attention, I was learning that spotting hawks from the Cessna through my 10 x 40 lenses was only very difficult. So, above the rolling plains I glassed the gaps between the approaching ridges, where I thought Amelia would go up into the Hill Country. As we swept in behind her, only a few feet higher than the rocky walls edging its inset agricultural pockets, Amelia's transmitter told us she was right below, and seconds later, at 60 m.p.h., Amelia reached the valley's head. That meant she would have to climb — and as the receiver told us, without losing speed Amelia pumped herself up over the crenelated rim of the Edwards highlands, dead on course toward the setting sun.

She had drawn Vose and me over another coast. Higher now, and in our speed-eating figure-S formation too far back to visually make her out, I told George that Amelia's signals were wavering so much I thought she'd soon be going to roost.

"Best to fill up now, so we can follow her in," he said, reaching for his maps. Seventy miles on, the Frio River's cypress-covered banks broadened enough, next to the town of Leakey, for a neat little landing strip, and since by then in spite of our slow flight maneuvers we had moved 15 miles ahead of Amelia, Vose cut the throttle and circled back. The landing field was deserted, and just before our wheels touched I saw the unmistakable slither of a long silver coachwhip as it slid off the runway. Head raised like a fast-moving cobra, it was pale, with an unmarked back and sides. At the coast, coachwhips were dark or broadly banded, but here in the plateau country their steely dorsums mimicked the chalky soil.

For me, that meant something. Amelia was really on her way and, subtle as the landscape's changes were, it was clear she had carried us into a different biotic zone. Binder, of course, had other concerns. Before our propeller coughed to a stop he was out of the plane and looking for a phone. Fortunately, there was one.

While he caucused with his secretary, Vose got down to business. "Immaterial," he snorted in Bob's direction, spreading his aerial charts on the edge of the asphalt and tracing our route with a red marker. "Be dark soon; got to figure where we'll bunk."

Stiff from sitting, I sprawled on the grass and let him decide. The land was as quiet as it had been at Armstrong's, and I could hear the soft step, crop, and munch of a herd of Spanish goats that tiptoed past us, filtering through the big white bee boxes that ran like rows of tombstones under the live oaks. The bees didn't bother the goats, who grazed right through the clover song hummed by the squadrons of yellow-brown workers who droned from flower to flower between the nannys' slow-moving legs.

Then a pickup appeared, driven by a cheery teenager who, after George had run through a brief explanation of radio-telemetry, replied that of course she knew what peregrines were, they were those big bird-things with yellow eyes that flew at night. Then she unlocked the gas pump, and 40 gallons later we were once more under way.

But this time it was different. Bob was still on board, reluctantly, but Scott Ward, the girls at the Antlers — even Jen — were half the state behind us, and I knew that Vose was right. All that was immaterial. Now there was only Amelia.

10.

The Prairie

Five hundred feet below my window, between overhanging limestone stained dark by runoff, ran a thread of bright water.

"Beginning of the Frio," noted George. "Next stream's the Nueces." To our right, a pair of intersecting valleys were lined with pale foliage. Dark-leaved oaks filled most of the canyons, but the pastel colors in these glens belonged to bigtooth maple, little walnut, and velvet ash. Collectively, they formed Lost Maples, a state preserve whose trees were called lost because they were remnants of the montane forest that flourished here in cooler, wetter times. With the first November frosts, most of the maples turn orange, some turn red, and a few go livid maroon. For a couple of weeks the woods look like Vermont, except that the trees are shorter and no lines of cars idle through the groves since the maples are back in the canyons where you have to walk a mile or two to see them.

But that happens on the fallow side of the year, under a dimming sun unlike this setting orb that, now approaching its summer zenith, made it almost impossible to look ahead through the windscreen. I was wondering how Amelia's immense light-gathering pupils could steer her into that glare, when her signal abruptly faded. Hearing it diminish on his own headset, George leaned us into a shallow turn, and sure enough, on the far side of the circle Amelia's pulse came back. The next time around it hadn't moved: somewhere along the river below she had gone down for the night.

"Let's have a look," Vose proposed happily, now that there was some real flying to do.

Only the canyons' eastern rims caught the setting sun, and below that lip their depths were dark, blocking Amelia's line-of-sight transmitter. Didn't matter to Vose, who dropped us below the illuminated eastern wall that defined our course up the Nueces gorge. Just off the rocks, there was no room for error, and back on the rear bench I could feel Binder start to fidget, but before either of us could speak, we had her.

Twenty feet down from the edge of the

plateau a dead juniper clutched the bluff with a snarl of desperate roots, and though we could see nothing in the shadows, from it came a strong signal. Triumphantly, George pulled up, then leveled out and motioned for his chart — onto which, next to the proper bend among the Frio's convoluted curves he inked a precise blue dot.

After so many nights following Amelia home to scrubby, thorn-brush roosts on the dead flat coastal plains, it was odd to find her bedded here, 2,000 feet up on a stark limestone precipice. But the home of her heart was high, and I guessed that from now on her roosts would be stony headlands and riverine bluffs. For it seemed that Amelia had not arrived on the Frio by accident. All day she had pressed into the wind, and only when she reached these stony cliffs had she folded in the long blades that had rowed her here and dropped decisively into that bare juniper. It was a spot, I told George, where perhaps she'd slept before, in previous springs or autumns, passing into or out of the terrain of her other life.

As the sun vanished and Vose climbed us slowly away from the river, we met an almost full moon as it, too, rose over the eastern horizon, and I remembered how, a month ago, searching for these hawks I had followed its silvered path across the flooded laguna; now it lit our way above dark upland canyons toward the town of Kerrville. As George radioed for landing clearance, in the calm air I ran the figures. Since going inland at Baffin Bay, despite an increasingly stiff northwest wind, Amelia had flown 258 miles. Even with her reconnoiter pause south of Poteet, that meant she had covered an average of more than 40 of those miles every hour.

It had been a long day for the three of us, too, and at the Save Inn, 5 a.m. came early. Wound up with the flight and unable to sleep, Bob had insisted on watching every bit of two late movies, which left him groggy, and since George wouldn't crank up his internal motor without a jump-start of coffee, in the airport's loaner sedan we straggled around Kerrville's pre-dawn streets in search of an open breakfast place.

It was a big mistake. When we found a café it was so slow I was on the brink of provoking trouble just to get us moving, but George, who prevailed, thought we ought to calm down because experience had shown that except to feed near their roosts peregrines never got going before ten. It was past 7 when the fuel truck rolled out to the Skyhawk, 7:30 when we turned on to the runway, and after 8 when I clicked on the receiver over the river.

Amelia wasn't there.

It was a gray day, and we rose toward the low clouds listening dismally all the way around the compass. It takes a long time to gain altitude with four small Lycoming cylinders aiming at 150 horsepower, and for half an hour all I had was static. Vose frowned patience at me and held his climb. The white hand of the altimeter was almost to 8,500 when, very faintly, I got a beep. George didn't hear it at all, and it didn't repeat as we finished our circle, but it had come from a bit north of our previous heading. The direction was hard to tell, though, because a solitary pulse like that could have arrived coincidentally while we were at the northernmost point on our loop; without a confirmational blip the only choice was to go higher.

As we lifted through the upper layer of stratus, the canyonlands disappeared below, and with it went our tenuous sense of connection to Amelia. Now we were just flying. Droning ever-higher orbits, the three of us and the plane were all that was concrete; our living, burning peregrine had once more become an illusion.

At 9,500 feet George looked over, and I pointed north-northwest. It was the only clue we had.

Have we got this bird's route or what, Bob wanted to know. I told him, sort of, but an hour later I hadn't heard a thing. Sheep and goat country was far behind, and through breaks in the clouds I could see flat red dirt: the beginning of the Great Plains.

George was convinced Amelia had simply left before daylight and built up a long head start, and since she couldn't outfly the plane's higher speed, because we knew her general direction of travel it was only a matter of time until we overtook her. To let our antennas sweep as much radio range as possible we had to be high, though, making big listening loops every half hour.

By midday the clouds had cleared, and squeezing everything out of '469, George had managed to climb another 2,000 feet. It was nearly as high as we could go, and over 2 miles up, if we were going to locate her it would be from here, inching across a world that had now become mostly sky.

By now the land had grown more civilized, sequined in the heat with the small shimmers of metal-sided barns and commercial buildings. That meant we were approaching Interstate 20, a four-lane slab running between Dallas and El Paso, whose midpoint is Abilene, a flat, brown-brick town. Just our side of it lay Abilene Regional, the sight of whose runways brought Binder scooting into the gap between our front seats, adamantly demanding down and out. It was obvious we had simply lost our peregrine, he insisted, and when Vose shook his head, Binder looked at me with amazement. I knew what he was thinking: George and I had become so possessed by the momentum of our mission that, like Kesey-esque gypsies, we were about to haul him farther and farther across the skies of West Texas, while his law firm, neglected three days at this point, slipped toward unanswered interrogatories and lost cases.

That ran headlong into Vose's strategy. George wasn't inclined to give up our hard-won altitude for a trip down to Abilene Regional, so it became a stalemate, with terse words on both sides.

Finally, I voted with Bob. Even counting the hour-long climb it would take to get back to our radio vantage point, Vose and I would be better off on our own.

A side-slipping arc out to the west brought us down on a steep approach and into a harsh, bouncy landing, after which George wouldn't take the time to taxi a single yard farther. From my door Binder clambered down on an outlying runway apron, where Vose solemnly shook his hand. Then, before Binder had made it 100 feet, George had us back in the air, on full climb.

But it was an empty sky we inched up into. No aircraft, not a vulture or soaring red-tailed hawk, certainly not our tiny wisp of a peregrine, broke the empty span of blue. As the horizon dropped lower, George started another big listening turn, this one 50 miles in diameter. Yet as we came back around that long, climbing spiral, our earphones were still empty.

Vose raised his eyebrows, looking at me for instructions. I had no idea; I thought he'd know what to do.

Then it hit me: Vose wasn't nearly as competent at radio-telemetry as he had appeared. And once we were off the coastal flats and away from the falcon trapping, I wasn't competent at one single thing about this endeavor. Each relying on the other's presumed know-how, George and I had blundered along this far on confidence and high spirits. But with 20,000 square miles of West Texas lying vacant beneath our wings, we sat helplessly, side by side, no longer truly in pursuit of anything, not knowing whether we were ahead of Amelia or behind her, or even if our belief that she would travel for days along the same heading held even the slightest validity. Amelia had kept to that course yesterday, but we had been absurdly cocky to presume it was her only option, much less that it would take her to Alaska. She could go anywhere, I said, and at last Vose agreed that we had no idea where she'd gone.

We were nearly to New Mexico before either of us spoke.

"We've been averaging over a hundred," George said. "Even" — he gave me a bad look — "with that useless stop in Abilene. Our old bird couldn't have gotten this far unless she left at midnight."

A rising crosswind had shoved us up to the Texas Panhandle, and turning back meant crabbing against its southeasterly flow. With his right pedal Vose fed the tail some rudder and bucked the Cessna's nose around into the wind. Then, as our antennas swept the southern horizon, there she was.

Borne on that big breeze, Amelia was headed right for us, and with a sly-dog grin Vose glanced over. I was silent; I had believed we'd never find her.

But just as George had figured, Amelia was still on course. She had simply fallen behind, though we couldn't have overshot her signal on this open tableland. Then I realized what might have happened. Amelia could have been down in one of the canyons at dawn. In that case, blocked by rocks and vegetation, this morning's single beep would have been only a ricochet, faintly deflected from the scarps, that I had taken for a far-off flying pulse. At the coast, peregrines liked to drink early, and Amelia's antenna might have been dunked in a riffle; one of the Padre falcons had gone wading in the laguna, submerging its radio wire, which temporarily cut off its signal.

But all that mattered was that we had Amelia back, and happily we flew on past, headed for a small municipal field where we could listen to her progress from the ground.

WELCOME TO CROSBYTON, CROSBY-

TON SEED said the little sign flapping from a huge rain gauge outside Crosbyton's flight office. It was a matter of priorities. A foot and a half long and an optimistic 8 inches deep, the gauge held only dried June bugs, a symbol of the parched sorghum fields that spread for miles in every direction, seed heads bowed northward by the hot wind. Leaning south, I edged into the lee of a hangar, thinking what a surprise it always was to step out of the plane into the invisible wall of the wind, because until you're on the ground you don't feel its pressure.

When you are airborne, turbulence kicks you up and down, but because you are part of a moving airstream you have no indication of how fast its flow is sliding you across the countryside except by looking down. Like swimming in a running river: there's not much sense of the current until you grab a rock and try to hang on.

As soon as its feet leave the ground a falcon doesn't feel the moving air either, and it never has to push against the breeze the way a sailboat, held by the water, struggles to beat its way into a gale. For a hawk or a plane, it takes the same effort to stay aloft with or against the wind; more ground just slides past beneath you, faster, if you're traveling in the same direction as the air mass.

Inside the hangar, still wearing its original aquamarine paint, sat a '56 Bel Air four-door: except for the color, my first car. Scrutinizing that perfect Chevy took a while, but behind it, squatted back on their tiny tail wheels, sat a pair of Cessna crop dusters. "Ag planes" to their pilots, they were pointy-nosed, single-seat craft that, except for the spray nozzles on the rear of their low wings, looked like World War II P-51 Mustangs. They were also deadly, though in a more subtle way; even from a distance they stank of Kelthane.

George, however, had already gotten down to swapping credentials.

"Want you to meet a real ag pilot," he called.

Standing next to one of the demi-Mustangs was Charlie Westfield. Vose patted him on the back. "Guy flies every day," he said. "Hasn't been higher than five hundred feet in fifteen years."

Westfield shook my hand. He also smelled like Kelthane. George was telling him about the Customs officers who had sliced '469's headliner, so I again leaned into the wind and made my way out to the Skyhawk, where I could hear Amelia, now so close her beeps almost leapt out of our speaker. With binoculars I scanned the sky for thin black wings. I saw nothing, but from even her limited altitude, I knew what she could see. Just over my horizon was the Caprock, a row of sandstone hills that marked the western edge of the tallgrass prairie.

Agriculture only went that far. Beyond was ranchland: high-country mesas and big dirt arroyos that ran all the way to the Rockies. It was a step closer to home and, like seacoast or tundra, those empty prairie sweeps, I felt, would have seemed richer to Amelia's eyes than the sterile sorghum country behind. And behind was where she was rapidly putting it, because after twenty minutes Amelia's signal had dimmed so much I had to haul Vose away from Westfield. An updraft bumped us into the air halfway down the runway, and as we climbed, turning with the wind, I could see sharp-edged cloud shadows racing us across the farm-to-market roads that gridded the dry fields. Vose pointed at them.

"Those country lanes are exactly a mile apart. Count how long it takes a shadow to get from one to the next and you'll know how much help we're getting from this breeze."

A hundred and three seconds: 40 m.p.h. That was a lot of current, and every time we swung right or left to check our transmitter bearing we would get a sideways, wallowing push, like a wave coming up under the rear of a sailboat running before the wind.

"Cessnas are bad in a quartering flow," explained Vose, correcting the yaw like you'd steer around a pothole. We spooned down a couple of the pop-top microwave stew cans we'd gotten from Crosbyton's vending machine, then set to work charting Amelia's path. We were right on course, but her signal was so weak that after a few minutes it looked like we might not be catching her at all.

George was excited. "We're doing a hundred and forty miles an hour on the ground and she's going almost the same. Look at this." Though '469 sat dead level in the air, our rate of climb read 1,000 feet per minute. Thirty miles ahead Amelia was getting the same big boost, and I imagined her exhilaration at flowing effortlessly up toward the clouds. Why else, after eleven hours on the wing, would she fly so high and so fast? Already past the Caprock, by now she would be approaching the Prairie Dog Town Fork of the Red River.

Ahead lay nothing but grass.

It was the High Plains, the elevated western half of the continent's great central shelf. Until a hundred years ago, as unbroken prairie, it stretched from the Gulf of Mexico to northern Saskatchewan. From the air back then, the specks of grazing Herefords below would have been pronghorns, mule deer, and bighorn sheep. This time of year there'd also have been bison, millions of ragged, winter-coat-shedding animals moving north with the spring. It was the last time you could have seen them like that, since by then their prairie sea had spawned the dreams of cattlemen — men whose vision of empire brought them, for the most part, only hard labor and bankruptcy, while they, in turn, brought ruin to the prairie.

By evening, Amelia's sprint had slowed, and we were weaving ever larger airspeed-eating Ss behind her. Ahead, a dark cleft pressed into the plains, and the instant she reached it Amelia's beeps went out. As we drew closer I could see that the rift deepened into a full-fledged canyon, and as we passed over its lip George announced, "Canadian River."

Below the Canadian's sandstone rim

three silver-braided channels veined its floodplain, and as we turned upstream Amelia's signal came back, strong and stationary.

Vose looked over. "That old gal knows what she's doing; got her overnight spots all picked out. Maybe has had for years."

I grinned to myself. This was what I'd hoped would happen: Vose, too, was falling under Amelia's spell, and I gave him a nudge and asked if he'd still shoot her for a matinee ticket.

Once Amelia was settled on this good roost, I was sure she wouldn't move again, so we headed toward Amarillo for the night. On the city's outskirts stretched a gravel airstrip, and we dipped a wing to check it out. It bore some checking, too, for in the field between it and the highway stood a vertical row of graffiti-sprayed automobiles, rear wheels raised to the sky, each car's buried nose sunk to its steering wheel in the earth. It was the Cadillac Ranch, helium heir Stanley Marsh and builder Doug Michels's monument to 1960s Detroit, and from the air its rusting tail-finned fuselages looked like junkyard arrows impaled in the plains.

"That's a shame," said Vose. "I love old cars. Got six; none of 'em cost more'n five hundred dollars. You've ridden in the best one."

It was a powder-blue, unintentionally low-riding '66 T-bird that George had picked me up in one day during a thunderstorm, and I remembered it because it had no driver's window.

"Hardly ever rains in Alpine," he'd explained, water streaming off his left elbow. "Still, I oughta get one of the others running, just in case."

To the east we could see the sun's last reflections shining from Amarillo's glassine office buildings, and above them a sky full of blinking aircraft lights. Vose jiggled the mike and tapped it on his steering yoke.

"See if you can get landing clearance on this thing."

Flight control was a mess, snarled by after-work student pilots herded through their touch-and-gos by a tower operator whose drawl I could make out only now and then. Our intermittently operative mike didn't give us much voice in the airfield's cacophony of transmitted positions, identifying numbers, and holding-pattern confirmation, but George didn't put much stock in the process anyhow. Serenely, he shouldered our way between a Cardinal trainer and a business jet, moved aside to let a Southwest Airlines flight go in, then, after a garbled go-ahead from the tower, dropped us onto the stacked-up runway.

Even more than usual, it was good to set foot on the asphalt. While Vose snugged the tie-downs, I went for a phone, and by the time he'd gassed up and funneled the usual quart of non-detergent 50 weight into '469's crankcase, the Budget lady was there with our car. We still had our little portable antennas, and along with our new scanner I stashed them in the backseat and beckoned to George.

Except for his own vehicles, Vose wasn't interested in driving, so I headed back out Interstate 40, making our normal convenience-store stop for a portable dinner, then turned north. Bordering something called Cal Farley's Boys Ranch and Girlstown, County Road 1061 crossed the Canadian not far from where Amelia had gone to roost, and across the lane's narrow verges spilled a soufflé of yellow coneflowers. Away from the road the prairie was even prettier, with tobosa and big bluestem that reached higher than our knees, and made our walking, once we'd slipped between the strands of barbed wire, like swishing through a coarse-bristled hairbrush.

Evening birdsong was everywhere, and I hated to pull on my earphones; but that was why we were here. Linked to me by coaxial cable running from the receiver hung around my neck, George followed down the slope, rotating one of the hand-gripped Yagis as high overhead as he could reach — about a foot more altitude than I could have provided. A trio of turkey vultures caught the light as they planed toward the setting sun, yet despite our hopes the sandstone bluffs held no upright peregrine silhouette.

Then, with skittering, moth-like flight, a dim shape fluttered up and perched soundlessly on the canyon's rim. It wasn't a peregrine, but we froze anyway. Through field glasses I made out the spindly legs and gray-flecked back of a burrowing owl, whose big circular face had rotated 180 degrees, completely back over its shoulder to stare rearward at us under startled, white-ruffed brows. Trying to make out our stationary shapes, the owl bobbed its head, indignant lemon eyes aglow, then swiveled its robot gaze back around and dived into the abyss. It was gone so quickly I hardly saw it leave, but in its place, across the valley flew the winnowing shadow I knew so well. Tight-shouldered and advancing only with flicks of its long black wing tips, a peregrine swept over the river, flared at something moving on the ground, and then, ephemeral as the owl, merged silently into the Canadian's downstream shadows.

George was facing the other way and I almost pulled him over, stumbling up to the rim. He hadn't seen a thing, and the river was dark below, so I waved the antenna myself: out over the canyon, back toward the prairie, up at the sky. Watching the receiverdials, Vose listened and shook his head.

"Must not've been Amelia; maybe they all come through here."

It was nearly dark when, still linked by our umbilical cable, Vose and I quit listening and trudged back toward the road. Our path led through a stationary sea of rounded hills, and though its waves were frozen, across all that twilit prairie nothing was still, for with every puff of air the tobosa shook its shaggy coat, flinging out droplets of white-tailed lark sparrows that flipped downwind and dived back into its mane.

George observed that so much prey might have been why Amelia was here, but that wasn't the case. Flashing their pale outer retrices, the lark sparrows looked like obvious snacks for a falcon, but they were aloft only for seconds, and with a thicket of prairie grass a wingbeat below they could afford to be boldly colored, offering the snowy edges of their tails as beacons to companions following behind.

It was the same with white-tailed deer, I said, animals that flashed their cottony-undersurfaced tails to lead fawns away from danger. Raising such a showy target in front of, say, a pursuing puma had originally seemed like a bad strategy to me. Then I saw what actually happens. Deer don't die from being chased, at least not for more than a few yards; they get ambushed. Once white-tails are up and running they are OK; the priority is to show their offspring, who might panic and find themselves boxed in by trees or rocks, an escape route. I was illustrating this to Vose by waving a deer-tail-mimicking hand somewhere in the region of my coccyx when all I could see was blazing light.

"What you boys up to?" came the quiet, hard voice. Then the light went off, and from the driver's seat of a beacon-topped, refrigerator-white Ford a flick of meaty fingers motioned us over. Farley, or one of his boys or girls, must have spotted us.

Sheriff McAllister was young, with hefty, clean-shaven cheeks. That and his name tag were all we could see of him, and rubbing his eyes from the police Q-beam, George squinted down at the patrol car's window. From his height it was a long way.

"Son," he rumbled, "we're not boys. We're radio-telemetry tracking a migrating peregrine falcon. Which has gone to roost in these river bluffs." He paused. "For the night."

"We understand y'all got some kind of radar out here," McAllister's deputy said in a nasal twang, advancing around the truck with his flashlight trained on Vose's little antenna. "Got any ID?"

While our driver's licenses were being run through the Department of Public Safety's computer, George and I showed the two of them our equipment. It didn't cut any butter with McAllister, but since the Army evidently hadn't filed its charges with the D.P.S. he had to let us go. But not without a warning — the first thing, I hissed to Vose, they teach in peace officer school.

"You boys be careful," McAllister told us. "Mind whose land you're walking on."

An hour before dawn, Vose and I lifted

off from Amarillo and turned west, sure that Amelia would still be on her roost along the Canadian. But by the time we reached the river she was gone. Gone so long that even from nearly a mile up there was no signal. In disbelief we looked at each other, then without a word George banked north onto yesterday's course.

It was impossible to feel bad for long, though, because from the air a cloudless May dawn above the Great Plains is something to behold. Without preliminary graying the Skyhawk was suddenly in full sun, and ten minutes later the plains' olive shadows brightened into a luminous absinthe that for the first time let me see why Vose so often slumped in boredom on the ground. Why he never took the wheel of our airport loaner cars, hardly gazed out their windows, and why his pale-blue eyes lit with interest only when '469 first coughed and spun her propeller.

Nothing on the ground compared with what we could see from above, and as the pearl-hued dome of our sky turned blue as cornflowers, below, the plains crept toward us in the still air, sprigged with tall white columnar silos that thrust up like metallic Stonehenges. But the grain elevators were not just symbols of technology: they were signs of the prairie's process of decline.

Those early cattlemen didn't bring ruin to the plains by killing the buffalo and fencing in the Comanche. Unknowingly, they attacked the grassland itself, at its roots, and grass lives by its roots. Among flowering plants, which include grasses, there are two principal strategies. One is to live but a single season, pinning your species' genetic hopes on the next generation and devoting most of your resources to seeds, flung far and wide in search of new ground. That approach tends to make a plant a bad-times specialist — an opportunistic exploiter of upheavals, inclined to colonize disturbed, freshly opened soil.

The other option is to dig in. Be perennial. With roots thrust deep into the earth, a different style of grass can survive the fire, frost, and drought that characterize the plains. Those species can't reproduce very fast, since budding from underground rhizomes or slowly spreading surface runners takes a long time — but they can hold the soil and, with the decay of generations, build more of it.

Throughout post-Miocene botanical evolution, newly sprouted grasslands have held annuals; mature ones, perennials. The original American prairie was perennial.

With a tall-, mid-, or short-grass cover depending on how far west you went, the prairie's springy mat of roots kept it alive. When the bison surged through in spring and fall, they chewed the standing stems to nubs, but they did so for only a few weeks. When the forage was gone they moved on. Having hoarded most of their energy safely beneath the surface, the roots could then spring back, pushing up new shoots almost before the last hoof had tramped over the horizon.

Cattle were different. They never left. They weren't as voracious as bison, but on the High Plains, where the sparse soil barely iced a limestone bedrock, they walked the prairie to death. Even spread out on the great ranges, their hooves steadily compacted that thin vegetative veneer, gradually pinching off the grasses' root mat against the underlying stone. And without cover, the sparse humus — the ages-built flesh of a lean land — dried up and blew away. Or, exposed to the region's seasonal deluges, a good bit washed downstream into the Gulf of Mexico. Throughout the plains, that left a sterile matrix of semi-bonded pebbles, known as desert pavement, that now supports no more than a balding remnant of the bluestem and Indian grass, a half-dozen gramas, switch grass, tobosa, and the intricate aggregation of low-level forbs that once flourished on these plains.

Wondering what was left of the grass, I had asked University of Kansas prairie ecologist Hillary Loring how much original prairie remained.

She made a face.

"One percent. Or less: none of the tallgrass is left on the plains."

"Where then?"

"Gully banks. Hillsides. Any place too steep for a plow. After the deep, level soil was plowed, all that was left were marginal, sloping little sites that even the old-timers tell me shouldn't have been opened."

Flying north along the hundredth meridian, the approximate ecological border between the eastern tallgrass prairie and the arid short-grass plains to the west, I could see that off the Skyhawk's right wingtip, only commercial competitor grasses were now allowed to grow. The minuscule seed heads of the original prairie perennials are not commercial; what is worth harvesting are the swollen seeds of the domesticated annuals. Their carbohydrate-filled kernels have fueled every human expansion, but to grow them you have to rip up the land, because corn, wheat, rice, and oats are natural recolonizers of the terrestrial wounds left by floods and fires; to provide their seeds that sort of disturbed terrain means wreaking similar small, man-made disasters all over the prairie. Often, two or three times a year.

The big diesel plows that finally broke the plains not only harrowed under its forest of grass, they also robbed the prairie of its most important biological stronghold — millions of tiny aquatic refuges richer even than the tallgrass. Before the coming of motorized agriculture, a CAT scan through the Great Plains' mantle of bluestem would have shown anything but today's smooth tableland. Beneath its stems most of the land looked like a golf ball. A slightly defective one, because instead of being circular, its indentations were as shapeless as amoebas. But they were often regularly spaced, evenly pocking the face of the plains, and during wet weather they filled with rainwater to form potholes ranging from the diameter of a washtub to that of a municipal swimming pool.

Almost none of the ponds lasted all year, but during most springtimes each one dented into the prairie a small oasis, contributing a few circumferential feet of land–water interface to a cumulative total of thousands of miles of forb- and reed-lined shore. Home to larval insects, crayfish, frogs, and salamanders, even the littlest ponds served as nest sites for a pair of stilts or avocets, rails or teal, all breeding so rapidly in the mid-continental spring that the hawks and falcons, with their chicks in prairie nests, waxed fat on the shallow ponds' surfeit.

When the land dried out, the biggest ponds held the life-sustaining moisture of buried mud during most of the summer, usually filling again in autumn, about the time the arctic migrants swept back in. No more than a tenth of a percent of that pothole prairie — just 200 acres of the third of Texas once occupied by those grassland tarns — remains, and the reason is agriculture's constant reopening of the food-annuals' exposed seedbeds. First oxen, then tractors dragging plows and discs and harrows all scoured dirt from the prairie's high spots and pushed it into the low places, season after season, gradually combing away the Great Plains' ponds, leaving the land smoothed and dry and empty.

But to get as much of those food-annuals as we consume today requires more, I told Vose. It means feeding those new man-made grasses with the second-hand energy of petroleum-based fertilizers, and then, like Charlie Westfield, protecting the grain with biocides that steadily accumulate in the soil. American agriculture now uses nine times the pesticides it dispersed when peregrines went into their post-war decline, yet it loses nearly twice as large a percentage of its crop to chemical-resistant insects.

George gazed down from his window.

"Those grain fields always looked kinda nice to me. Enjoyed cruising over them." He looked at me. "Used to, anyway."

11.

La Junta

Ahead were the verdant swaths of the Rita Blanca and Kiowa National Grasslands. Their tall, native prairie was interspersed with agricultural inholdings, but in the refuges' thick-stemmed, old-grass sections, silver potholes shimmered back the sun. Then, with still no trace of Amelia, we were out of Texas and into New Mexico.

George put us into one of his long listening loops while I focused on the receiver's electronic hum, closed my eyes, and tried to will in a far-off pulse. After ten minutes I gave up, blinked, and looked down. For a second it was unclear whether I was still imagining the nineteenth-century plains, because below, bison flowed up the side of a ridge.

They were the Folsom herd, a sleepy-eyed bunch pastured here as a tourist attraction. Through binoculars I could see the herd bull treading ponderously through a froth of lavender verbena. He was impressive, but only half the size of his predecessors, whose fossilized remains were the reason he had been brought here. Literally a crossroads, the ghosted-out town fronting his pasture had only a few buildings, the largest of which was the shell of the Hotel Folsom, built in 1868. Long closed, with arched, glassless windows looking out on the sage, it was barely more than a decade old when a black cowpoke named George McKenzie discovered a heap of gigantic bones 12 miles north of town.

For nearly fifty years McKenzie's find lay untouched; then the Denver Museum of Natural History sent a team to excavate the site, which put Folsom forever on the paleontological map. From between the ribs of skeletons of big, long-horned *Bison americanus,* the Denver bone-hunters dug up roughly worked lithic projectiles that were hammered, 12,000 years ago, out of stone cores carried here from the Rockies. At the time, they were the earliest evidence of humans in the New World.

A few Folsom points, dug from those skeletons, still lay in the dusty cabinets of the town's only operating business — a storefront museum — across the street from the hotel. Along with heavy, lethal-looking axes, each of whose flaked-off serrations was the size of a half-dollar, the points were appropriate to their targets. The late Pleistocene fauna the Folsom hunters found here — ground sloths the size of two grizzlies, woolly mammoths, Galapagos-size land tortoises, and, mostly, huge bison — took massive spears to kill: stabbing spears with giant stone tips that made them too heavy to throw.

Like wolves, the people who used those weighty lances followed the herds across a savannah wetter and greener than the prairie the pioneers saw, a land punctuated by the still intermittently steaming volcanoes Raton, Clayton, and Capulin. Now Capulin's dormant cone thrusts a mile up from the plains, concealing the valley where Butch and Sundance hid out a century before George and I crept by overhead.

As I gazed down, searching Capulin's wooded flanks for their Hole in the Wall, my seat slammed up beneath me like a too-powerful elevator, shoving the Skyhawk 50 feet in the direction of the overlying cirrus. It was an orographic current, a heavy flow of air sent skyward by Capulin's steep sides. As we rode its gentler following wave above the crater's rim I had time both to settle a little and realize that this was a site sure to draw birds of prey, so with '469 soaring on toward 12,000 feet I listened hard, then slumped in disappointment. Amelia wasn't here.

But the ravens were, a swirl of black specks surfing the swells of moving air. Camped, years before on these slopes, I'd watched the ravens wake in the lower ponderosas, shake out their sooty feathers, and as the sun soaked into the plains, sending the first warm drafts upslope, cup their feathered arms and buoy away into the sky. By the time the first of the flock had passed the rim, at 8,100 feet, there was no way to miss the fact that every one of them was wrapped in euphoria.

Croaking for all they were worth, one raven after another would pull out of the group, fold its wings, and nose-dive off the wave's invisible breaking crest. As soon as the first trailblazer went, he became the goat. Tumbling in roller-coaster pursuit, his companions would try for a tag, but in the waves of the air the fugitive had the advantage. Beak thrust straight at the earth, he'd hurtle past the volcano's rim, then instantly cut back into the strongest part of its updraft and crow triumphantly as he lofted past his falling pursuers. After that, full of himself as ravens always are, the victor floated back downslope, his bright, ebony eyes scanning me for signs of a sandwich.

As Vose and I bumped through Capulin's air pockets, we spotted the tiny dot of a solitary hawk above the ravens. Through binoculars I could see that it was a migrating rough-leg, beating stoically along the volcano's face, also looking for lift on its way north. Then Capulin was past, and 90 miles to the west the horizon was sheared by the Sangre de Cristos. I hadn't thought of those peaks in years, but against the cobalt sky their icy, serrated teeth looked exactly the way I'd seen them first, as not much more than a toddler, in the company of Winnie Rockwood — a down-on-her-luck newspaperwoman whom my mother had more or less permanently handed me off to after finding that raising a newborn was far too demanding.

It was the luckiest thing that ever happened to me. After that, for years Rocky and I had meandered here and there — coming upon, somewhere in the course of our travels, a diorama of a taxidermied grizzly standing against a painted backdrop of the same white-capped peaks, which was so realistic that on hiking up into the real

Sangre de Cristos I'd been terrified, scrutinizing every fallen fir and shadowy gully for hidden bears.

Now, Vose's charts said the stony wall ahead was more than 12,000 feet — a barrier, still locked in winter snow, that we could not cross. Peregrines were supposed to slip through the passes between those high peaks, and I winced to think that by now Amelia might have already propelled herself up one of those porcelain valleys. Still, our headsets filled with the buzz of her absent signal, we pressed on toward that jagged horizon. With every mile it rose and became more craggy, and finally we could go no farther: in front of '469's nose an icy wall of granite cut off the sky.

"Formidable mountains," observed Vose. "But our old bird's got too much sense to go up in there."

He shook his head at the barren rock, then angled steeply off to the right, paralleling the face of the range.

"Wouldn't worry," he added. "She's still out on the prairie. Like the eagle."

Partnered with a U.S. Fish and Wildlife agent, George had once radio-monitored a young golden eagle. Trapped in Texas as a potential sheep killer, the eagle was taken to the northwestern state line and released.

Partly to make sure it didn't return, Vose was hired to follow it. But not being a real migrant, the eagle had no destination, and for days, he told me, it had wandered aimlessly along the Rockies' foothills. Now, without Amelia, it appeared that we were going to do the same.

It was that or go home.

Colorado was coming up and, since '469's engine stutter had worsened, we slanted down to the Cimarron River and turned toward the airfield our maps indicated at La Junta. I was looking for a town, but on the river's bank what I spotted was a castle. A stronghold made of reddish mud, it looked a bit like Santa Fe, with thick, earthen walls enclosing round-contoured buildings whose roof beams protruded in rows from their upper stories, though a pair of raised parapets ringed with firing slots added a medieval element.

"As though an 'air built castle' had dropped to earth" was how trader Matthew C. Field had described it in 1840, for we had come to Charles and William Bent's old trading post. "Built of the simple prairie soil, made to hold together by a rude mixture with straw and the prairie grass itself, and constructed with all the defensive capacities of a complete fortification," Field went on, and right he was, for the post looked as imposing as it must have in its heyday.

That heyday wasn't long, covering only the brief period when Anglo trappers and merchants, Native Americans, and northern Mexicans lived and traded together throughout the Southwest. Arriving on the upper Arkansas in the 1820s, the Bent brothers and their partner, Ceran St. Vrain, organized a sprawling fair and trading congregation that took place every winter, when the trappers had returned with their furs from the high country and the Plains tribes grew sedentary since the dead grass wouldn't feed their ponies well enough to let them follow the bison. Pulling in buffalo robes from the grassland Kiowa and Arapaho, beaver pelts from mountain men, and Mexican silver from Taos, the Bents swapped eastern trade goods that had been wagon-trained out from St. Louis.

After Will Bent married Owl Woman, the daughter of a Cheyenne priest, and managed to establish peace with her nation's traditional Apache foes, the Santa Fe Trail shifted its northern branch to this northern route along the Cimarron, where the Bents' place was its midway stop.

But Little White Man, as the Cheyenne called Will, could only hold his trading confederation together as long as there was open space, and grass, and meat to hunt. When Stephen Kearney's army took over Bent's fort on its way to the Mexican War, a path was opened for gold seekers, then settlers, who streamed up the Cimarron. On their way across the plains they had fouled water holes, burned all the scarce, open-country wood in their campfires, and driven off the bison, which when the wind was right could detect a human scent 9 or 10 miles away.

By 1847 warfare against the settlers' incursions had begun in earnest, but by then the tide of Europeans had grown too great to stem, for with them had come an even deadlier foe. Like the smallpox that had arrived with the first wagons up the Arkansas River ten years earlier, cholera was an alien microbe. With no antibodies to counter its intestinal assault, the disease swept through the trading parties camped outside Bent's post, then followed the tribes back to their villages. Groups fled to other encampments, but no one thought to stay away from the afflicted, and 80 percent of those exposed eventually died. In 1849 Bent abandoned his Plains diplomacy, set fire to the burnable part of his post, and headed back east. Owl Woman stayed behind.

From the air I pictured the last tepees pitched on the meadow just downstream from the fort, filled with final trading delegations wanting, before the inevitable, to splurge on clothes, beads, mirrors from Philadelphia and even Europe, rum from the West Indies. Then a vibration hit the plane — a rattling shake embedded in a shriek that penetrated even the sonic doughnuts covering my ears.

"Son of a bitch!" said George, twisting right and left in search of the sound. Then he pointed. Above the old adobe walls a blue-black F-14 — a swept-wing trainer from the Air Force Academy at Colorado Springs — ripped along the river, bowing cottonwoods in its wake.

"Boy's got no business down so low," Vose snapped, climbing '469 away from the scene, which let us see, ahead, La Junta. That was easy, because even from 15 miles out the airport's pavement filled the horizon. For once the Skyhawk's radio was working, and when I called for a designated runway the operator chuckled and told us to take our pick. There were three, each so wide we could have landed '469 crosswise, and we set down as soon as we saw concrete. From the ground it wasn't clear whether we were aligned or not since I couldn't see all the way to the grass, but according to Vose the airfield was a World War II bomber base, and at 4,400 feet of elevation, especially with student pilots, the piston-prop B-25s of that era needed every square yard of those massive runways.

Somewhere between Bent's old fort and the F-14, La Junta hung suspended in time. Toy-like by contemporary measure, hangar after wind-swept hangar slid past our taxiing windows, the open sheds' broad, still partly shingled roofs supported by massive wooden rafters. At the end of the row stood a big clapboard office, behind whose flight desk Leo Roeske poured George and me cups of coffee. A former flight officer here, Roeske was flanked by five or six airfield duffers and a couple of student pilots, all alerted by the radioed prospect of company and assembled to meet us.

Leo was the one I'd talked to, and he told us we were warming up in the old dental building. It seemed large for a dental office, but Roske said that during the war more than 8,000 people had had their teeth done here. Behind its long back porch, the crumbling walls of administration buildings stretched away to an officers' golf course, which still grew a crisp Bermuda lawn.

Roeske waved at it. "This is only half the picture. We used to have a mirror image: south side of the field was a whole separate airbase, exactly the same as this, except for the golf course. It was for blacks."

The oldest duffer was Willard Lantz, who had worked on this side of the field during the war but was more interested in telling us about the two John Wayne movies that had been shot there.

"After they were done filming, the Duke came back for a visit," he said. "Nicest guy you could imagine."

There was no question who the current Duke was. Chunky but hardly older than his pupils, instructor Dennis Robinson had no more than to clear his throat to draw every eye and ear. He asked Vose where we were headed, which was all the encouragement George needed to beckon everyone out to inspect the Skyhawk as he launched into Amelia's story.

Robinson, who was actually wearing a black leather flight jacket with a pale blue scarf tucked into its pocket, had more important things to attend to, and I went along as he strode toward an outlying hangar. Inside, parked in the center of a space built for a squadron of B-52s, was an immaculate electric-blue Pitts. An open cockpit biplane with stubby double wings laced together with gleaming guy-wire, it was a craft for pulling g's. That jacket wasn't merely an affectation.

Robinson hauled a matching blue helmet off the cockpit's foam-padded roll bar, waggled the Pitts's aerobatic control stick, which thrust out of its floorboard, and said he'd be going up in a little while, if I wanted a ride.

But '469 needed to be oiled and refueled, and to have its timing fiddled with, and while that took place Lantz, Roeske, and Vose got into the war they'd all shared. Especially the Women's Army Service Pilots.

Late in the conflict, mounting casualties among male fliers had prompted scores of female aviators to be drafted into ferrying military planes and towing target craft for Army Air Corps trainees. But there was opposition to the program, and George had the lowdown.

"Those gals' flight training was the pet

project of General Hap Arnold's girlfriend, Jacqueline Cochran, and for all us flight instructors, getting to teach them was the all-time plum."

During June 1944, with five other Pennsylvania combat instructors, Vose had found himself about to pluck that plum, mustered out with orders to report to Sheppard Field, Texas, where the new WASPs were to be trained. Roeske and Lantz rolled their eyes.

"Yeah, but more and more pilots kept arriving," Vose went on, "and not just instructors. After a while we caught on that we were just pawns. We'd only been shipped to Sheppard so that when the WASP program came up before Congress the opposition could kill it by saying there were already more'n 500 experienced fliers in Texas."

Everyone groaned. Entirely in his element, George beamed, and for a moment I caught a glimmer of what in his heart of hearts Amelia — even me, for that matter — really meant to him. Without us he wouldn't have come to La Junta. And if he had arrived there in '469 on some ordinary cross-country jaunt, he'd have been just another airfield tourist; he wouldn't have been the long-distance, Charles Lindbergh sort of guy he was today.

Not that the new Duke was taking George's intrusion into his spotlight lightly; over our heads came the whine of an engine. From the dentists' back porch we all looked up. I couldn't pick out Robinson's scarf, but as he roared past there was no missing his bright blue helmet.

Five hundred yards from the flight office, he pulled the Pitts into a tail-stand and hung, for a moment, nose up from its prop. Then he stalled into a fluttering backward collapse, like a falcon-hit goose, and I silently gave thanks I hadn't taken him up on the ride.

That fall used up half the Pitts's altitude, which left Robinson just enough airspace to twist into a pair of horizontal barrel-rolls before he disappeared over the western horizon.

"Damn dramatic," acknowledged Vose. "Hope that boy doesn't hurt himself."

Then the old long-distance flier rolled the two of us out into a corner of those thousands of acres of unused concrete and took off for Denver.

Part Two

ON THE WING

We believed we had been made whole,
at least a small wing's width
of light in some bird's dreaming itself
back across the continent;
a part of that infinitesimal turning
that sends a hundred species skyward,
sends them home,
though they fly, sometimes, into air so cold
    it kills them;
into fields so stripped they'll starve among
    the weeds.
But isn't it instinct's greatest vision
to extend what's possible,
to risk not coming back . . .

Deborah Digges,
"Circadian Rhythms"

12.

Horseshoes

Whether or not Amelia had gone past Denver, its aviation repair hangars were where George and I had to go, because '469 was in trouble. Its engine had continued running roughly, but the larger problem was its brakes. The wheel on my side had briefly locked up back at La Junta, then broken its shoes free. Now Vose could reliably halt only the wheel beneath his side of the plane, which meant that on the ground we could only pivot in that direction. Touchdown at Front Range Airport went all right except for an extra long rollout, past two commercial jets waiting their turn on the runway, and a row of shut-down hangars. Then we got a taxi lane directive to turn right.

"Roger," George told flight control, "but we're going to need some other instructions. We can . . . well, sort of only make lefts."

Three taxiway lefts produced a right, and rotating our way around the field's network of concrete corridors eventually brought us to the repair apron. This time we didn't have to wait for Vose's stories to get a turnout.

Yet none of those who'd left their desks to watch our progress was able to help, because it was Saturday and airplane mechanics don't work Saturdays. I hadn't been keen on carrying the Skyhawk's spare asbestos brake pucks in the plane at all, but George had insisted on bringing them, along with an assortment of hydraulic lines and fittings he kept in a fruitcake pan wedged under his seat, and while he dismantled our disc caliper I borrowed the flight-office clunker and — slim hope that it was — went to look for Amelia.

If she was still somewhere on the prairie she'd be behind us now, so the place to listen would be back to the southeast. Highway 70 ran out that way, and earmuffed in my cucumber-colored headset, with our spare chrome-branched Yagi clamped to the car aerial, like some manic CB hacker I cruised the plains.

Yellow-throated devil's claw, prairie gentian, and ground cherry — all first described here by Lieutenant William Albert of Kearney's Mexican War Army — had burst up by the roadside, long stitcheries of snow fence seamed the pastures, and where cattle had overwalked their turf, spurge and knapweed took the place of bluestem and wheatgrass. In the valleys, alfalfa was getting under way. By Labor Day every stalk would be topped with a purple bloom; by Halloween, cut and rolled into shredded-wheat loaves the size of minivans, it would be stacked in pillowed hedgerows for winter.

Meanwhile, I stood on the picnic tables of every roadside turnoff, rotating the other small antenna overhead as onion trucks and mud-spattered horse trailers roared by, their drivers occasionally craning around for a longer look. Then, two hours out, my turnaround came at Punkin Center, which had a good name but was only a cluster of metal sheds where two country lanes crossed, although as I slowed for its intersection a gaudily black and white little bird caught my eye. Fluttering up and down just off the pavement, it wasn't engaged in one of those broken-wing ploys, but its flight didn't look quite normal either, so I stopped.

It was a male lark bunting, a species that in spring pours northward through the Rio Grande Valley. But with this one something was different. He looked the same, with aniline black and ivory wing bars that spread into pale shoulder pads when he flew. Yet from the top of a tilted fence-post he flapped slowly up to telephone-pole height, then splayed his wings and coasted back to the same perch. As the wind died, I heard his song, and then I knew.

My South Texas migrants had called to one another only in low whistle-coos, but this little guy wasn't still on the road. He was home, set up on his own patch of prairie turf and singing out his heart to defend it. Warbling and trilling, he again angled upward in a shallow climb that gave him plenty of time to show off his harlequin wings. At the apogee of each small trajectory, the bunting capped his rising aria with a couple of shrill whistles before he nosed over, dropping back toward the post on which he'd hardly settled before he went hovering up once more, singing all the way.

It was song-flight, that magical performance developed all over the world by small grassland birds. Among most song-birds, the males' springtime vocalizations claim territory in roughly the way that scent-marking does with dogs. Among birds, color and sound fill the role that smell occupies in the lives of most mammals, and commanding the highest singing post is the best way to saturate a stretch of land with the proprietary stamp of your voice.

In most places that means taking over a treetop post, but for Great Plains birds trees are an almost unknown resource, and to attract a female to their bit of turf, the males use their wings to create out of thin air their own elevated prosceniums, swirling and fluttering as they fill the sky with song.

It is an enchanting spectacle. In any opera, the vocalist can never be too far from his audience — in this case, a bird's mate as well as his potential rivals — so sky song is often repeated right in front of a listening human. What is most moving about it, though, is the singer's gaiety. Territorial though these performances are, the spontaneous joy of male song-flight is no projected human emotion, for only a genuinely ecstatic individual, bubbling with the vigor inspired by both his mate and the glory of a springtime morning, is compelled to fling himself aloft, hour after hour, pouring out his enthusiasm in crescendos of song.

With that in mind I looked for the recipient of all this gusto. But my bold little bunting's gray-brown hen was so well hidden, perhaps already sitting tight on a nest in the grass, that I couldn't find her. She must have been nearby, because when I walked up to his cedar post, her mate only flew higher, trilling longer and longer glissandos until I went back to the road and let him return to his perch.

Yet this morning's cheerful bunting song obscured a larger context. And, as is so often the case when our attention is captured by wild creatures, it was one of disaster. Attention has been focused on declining woodland birds whose tropical forest wintering sites are rapidly disappearing, and on the sudden fall in population of shorebirds who every year find their coastal feeding grounds further drained or built upon, yet birds of the prairies have suffered even greater losses. A U.S. Fish and Wildlife Service study on the eastern Great Plains found that between 1960 and 1987 bobolink populations fell 90 percent; those of grasshopper sparrows, 56 percent; savannah sparrows, almost 60 percent; and the lark and field sparrows that George and I had watched spurting out of Cal Farley's Panhandle hills, 54 percent.

They are gone mostly because they lived

in the most accessible of all North American ecosystems, and the one most desirable to mankind. The plains' shifting rainfall and uncertain populations of insects and seeding grasses always rewarded prairie-living birds whose wandering instincts could send them on to greener pastures. But now so much of the continent's grasslands are either farmed or covered with housing developments that even nomadic species like lark buntings are in decline because there are no longer new places left for them to go.

Nevertheless, even in its diminished state my springtime prairie was so wonderful I didn't want to go back to Denver. Amelia had to pass through sometime, so at a little civic park in Strasburg I pulled over for a final listening scan. Waving the antennas back and forth, I noticed that the enclosure's lawn was incongruously adorned with a bed of red, white, and blue nasturtiums surrounding a wooden monument that announced PACIFIC on one side and ATLANTIC on the other.

Between the oceans, a roofed sign informed visitors that here

A CONTINUOUS CHAIN OF RAILS FROM

ATLANTIC TO PACIFIC — LONG A VI-
SION OF PIONEER RAILROADERS AND
FRONTIER-TAMERS — FIRST BECAME A
REALITY AT 3 P.M. ON AUGUST 5,
1870.

The sign went on to describe how the
Kansas and Pacific rail-building crew
driving west from Kansas City, and a
Union Pacific crew, which was building
east from Denver, had lain a record $10^1/_4$
miles of track that day, vying for a barrel of
whiskey their foremen had set midway in
the final gap.

While I was reading, a tall, rancher-
looking couple pulled up carrying a
mortised hardwood box. They were Keith
and Janey Williamson. He was weathered
and reticent, but Janey was small-town
friendly.

"What about Ogden, Utah?" I asked
after we'd introduced ourselves. That famous railroad-linking golden spike?

"They *do* celebrate that over in Utah,"
Janey conceded. "Ceremonies and everything. But so do we. You ask Ms. Emma;
she's our historian."

In the Williamsons' wooden chest, which
Janey was unpacking, I could see rows of
routed stalls. They were for horseshoes.

Next to the railroad monument, a cyclone-fenced facility I had taken for tennis courts was actually a horseshoe stadium: eight lanes of play, paved walkways around sand-filled stake pits set below twelve tiers of covered bleachers and two center-court awards tables. It was the Wimbledon of horseshoes.

As president of the Colorado Horseshoe Association, Keith had come into town to get Center Court ready for the upcoming state tournament. But he was going to toss a couple first. I asked if he'd ever won the tournament.

"Nope," he said cheerfully. "But I run it."

At Ms. Emma's place I learned from tenaciously enthusiastic, chalky-haired Emma Michele that the Strasburg she served as permanent historian of was only the most recent of five previous Strasbourgs, starting with the medieval Strasbourg in Alsace-Lorraine, famed for its cathedral. Its Catholics were driven out early in the nineteenth century, she said, only to found, in the course of their long and much-persecuted travels, a Russian Strasbourg, another in Sweden, and yet another across the Atlantic in Pennsylvania.

There the promise of cheap land and government loans drew those German farmers — the hardworking, debt-paying settlers recruited by both the Burlington & Missouri and Santa Fe Railroads, who owned all the land for 50 miles on either side of their tracks — to colonize their vast acreage. Out on the plains, the Strasbourgians vindicated the railroads' social profiling by founding not one but two new Strasbourgs: Strasburg, North Dakota, and Strasburg, Colorado.

Ms. Emma didn't know how the spelling got changed, but she admitted that so far Strasburg, North Dakota, had done better — somewhat undeservedly, she felt — by cashing in on being the birthplace and boyhood home of Lawrence Welk. Nevertheless, under her guidance Strasburg, Colorado, was aiming at greater relevance by promoting itself as an alternative to Ogden, Utah. There was a lot more, but I had to get back to George, so I told Ms. Michele I'd try to publicize the truth about the railroads. She said not to forget the Williamsons' upcoming horseshoe tournament, and as I pulled out of her driveway the iron clang of a ringer came sailing up the road.

"That'n hit hard!" Janey called. "Like a chime."

★ ★ ★

Without parts, nothing much could be done for our motor, but by late afternoon Vose and a couple of volunteer brakemen had Skyhawk '469 once again able to stop straight. After having already listened for Amelia on the prairie, after takeoff I motioned George into a transmitter-scanning arc that carried us over the foothills to the west, past the Flatirons — huge slabs of granite geologically flipped up onto their sides, which had reminded the arriving fur traders, who first saw them from 90 miles out on the plains, of Paul Bunyan's pancake griddles. In California, rock faces like that would have held falcon aeries, ledges defended by resident peregrines who had fought for homesites so valuable that if either member of a pair had vanished, pressure for its reproductive niche would have drawn a replacement mate within days.

Not here. Jerry Craig of Colorado Fish and Wildlife had told me he was aware of only four breeding pairs of peregrines along the whole of the state's eastern mountain wall. Three other aeries were occupied by nonbreeders, but the rest of the Front Range — 300 miles of historically falcon-populated granite cliffs overlooking grassy plains filled with prey birds — was apparently without peregrines.

"Nobody else living there . . ." said Vose. "You'd think Amelia'd be up on those cliffs right now." A gust slammed my wingtip down, twisting the horizon 30 degrees, but George hardly noticed, letting the Skyhawk wallow its untended way through the flak of air bubbles that spun off the mountains. What he was thinking about was why these fragile little hawks wanted to go all the way to the Arctic. What was the matter with right here?

Probably nothing I said. But until the mid-twentieth century southern *anatum* race peregrines, as well as prairie falcons, nested all along the Rockies' eastern wall. That was one reason the slightly smaller tundra peregrines, passing through on their way north, had to keep going: no vacancies. I'd often watched West Texas peregrines blast off their nest ledges, screaming at anything from vulture to raven that entered their airspace, and trespass by an alien falcon meant real battle, sometimes to the death.

Still, Amelia's tundra race had probably once occupied these very cliffs. In the long eras of ice, when these stones stood as outposts of the perma-frost, the plains below would have been tundra: home to arctic peregrines that were later forced to follow the great glaciers north in retreat to their current range on the continent's polar rim. But peregrines are traditionalists about their nest sites, often occupying a ledge scrape that is the ancestral home of one member of a breeding pair — which means that even the most appealing new territories are slow to be colonized.

Here, I was glad of that. Glad that Amelia was headed for purer nesting turf, because even though federal regulations had largely eliminated DDT north of the Rio Grande, along the route Vose and I had flown so far agricultural chemicals seemed to be everywhere, even substances like Charlie Westfield's Kelthane — DDT with a single extra oxygen atom that metabolically breaks down into the same eggshell-thinning dichlorodiphenyldichloroethane (DDE).

The year before, Bill Satterfield had proposed to the Environmental Protection Agency a project examining the response of peregrine immunoglobulins to the generalized cellular insult set off by the organophosphate and petroleum distillates that have permeated the environment from thousands of sources. The difficulty of finding an uncontaminated wild adult from which to obtain baseline values was among the problems that had brought the project to a halt.

It wasn't that all the remaining peregrines were about to disappear, but with so many harmful elements at large it has become almost impossible, especially with creatures as biologically sensitive as birds of prey, to determine exactly which substance is doing what and to whom, and today linearly generated causalities like DDT-thinned falcon eggs are regarded almost nostalgically by researchers remembering the simpler early days of their field.

One of the pioneer detectives of chemical toxicology is Robert Risebrough. Hawk-eyed and sometimes talon-tongued, Risebrough is a professor of environmental biology at the University of Southern California. His Ph.D. is in organic chemistry, and like Riddle and Satterfield he's also a bird of prey man. Yet, as is typical in the field of environmental sleuthing, much of what he knows about poisoning in hawks and eagles was learned elsewhere. In perhaps his most famous case, pelicans were the subject.

The trail started in southern California. During the 1940s, when no permits were required to commercially produce organophosphates, the world's largest manufacturer set up operation outside Los Angeles. One of its waste products was sulfuric acid, saturated with DDT, which was dumped into the Pacific near Santa Catalina Island. Only a few birders seemed to notice, but within a decade Santa Catalina's considerable bald eagle population had vanished.

By 1961 environmental regulations required the plant's sulfuric acid to be trucked to inland landfills, but even though brown pelicans never went inland, the Channel Island pelican colonies were also slipping into decline. Risebrough found high levels of DDT in their shells and in the fatty membranes of their eggs. He ultimately traced that chemical spoor to some 200,000 tons of industrial waste buried high in the Santa Monica Mountains, where, with every rain, the soil leached toxicants into the upland gullies that fed the Los Angeles River, which funneled runoff straight out to the pelicans' offshore rookeries.

The landfills were sealed in 1970. By 1975 brown pelicans had again begun to reproduce, and by the late 1970s they had largely rebuilt their old colonies. If only it were all so simple, for beginning in 1981 captive-reared bald eagles were introduced into the now-ostensibly cleansed Channel Island habitat. Immediately the eagles became chemical sumps.

When I spoke to Risebrough during a later study conducted by California's Bureau of Land Management, I asked him where the second batch of Channel Island contamination had come from. It was the sort of question he addresses daily, and he took a deep breath at the magnitude of my naïveté.

"Nobody knows," he finally said. "In 1986 the last California Condor egg, which never hatched, was laid in the wild. The parent birds were being artificially fed — entirely on human-supplied beef — yet when we looked at that egg's membranous fat, it contained toxic levels of DDE."

Bent's Arapaho and Kiowa rose to mind — their dragoon-carried cholera, the enigmatic smallpox that had ravished them before the white man's mysterious measles and scarlet fever. In expiation for their unknown sins, the braves that brought home the fevers rolled in fire, burned their tepees, and slit their ponies' throats, and though the issue is seldom addressed in the stunningly complex publications of environmental scientists, to listen for long to the best of them brings one to the realization that, to a great extent, none of them entirely understand the problem either.

What they know is that no one truly has a handle on the impact of the free radicals, recombinant proteins, and thousands of newly forged molecular bondings inventoried every month in the warehouses of chemical laboratories, much less what is taking place in the third world's garage-based biocide industry, where exterminators whip up whatever organocholorine stew they believe will kill the locusts that threaten village crops without immediately sickening the local children.

George said he also thought it was a good thing that Amelia was probably a tundra falcon, since that ought to keep her pushing on past this mess. But she wouldn't be pushing on much today. Front Range's weather radar had shown us a giant amoeba of arctic air that had budded off an Alaskan storm and was oozing down through Alberta onto the northern plains.

Already its turbulence had set Vose to work with the plane, seesawing back and forth against the oncoming gusts that had ground our speed down to that of fast-lane traffic on I-25. Below, I watched us ease past tractor-trailer rigs, then lose out to a red dot sports car. George and I exchanged bemused glances. We were barely moving, we had nothing on the radio, yet our own trajectory — our will to just go on, riding the back of the northward-spreading spring, reveling in its ever-earlier dawns and steadily lengthening twilights — was as strong as ever. It was what Amelia, wherever she was, had given us.

13.

Reunion

From just over the Wyoming state line I picked out Cheyenne. It was small and orderly, with spruce trees surrounding red-brick houses, and from the air it looked generations older than the commercial sprawl of metro Denver.

At the edge of town was Warren Air Force Base. It was the first time I'd seen it, but for me Warren was an important place: among the earliest photographs of my parents together were a handful taken there. Not long married, outside their officers' cottage they'd stood smiling in the deep snow of 1943. You can't tell from the pictures, yet under her long wool coat, like a lot of the wives whose men were headed overseas, my mother was pregnant.

But my father's part of Burma was some of the last foreign soil the Japanese relinquished, long after V.J., and since by then he was torn with shrapnel and destined for years in military hospitals, I didn't really meet him until nearly five years later. By then it was too late for us to connect, even as friends, and after he finally came home we fought so often that the pictures of the two of them happy in that wartime snow together — with me there too, in warm embryonic sleep — were the best family images I had. As Vose cruised on, I stared down through the dusk looking for the spot where those sepia Kodaks might have been taken.

Among nomadic peoples, I told George, the site of one's conception was the only place a person's spirit could reenter the soul's dwelling place in the world below. Warren Air Force Base didn't suggest much of an underworld, but my life seemed so scattered that even from 1,000 feet up, the sight of those old officers' quarters filled me with the longing of every rootless psyche for some piece of inviolable home ground.

Vose thought that was all pretty silly, but — not that he'd ever wanted to settle down, have some woman he'd have to check in with all the time — he guessed that a person with enough imagination might see us as a couple of old gypsies ourselves. Then, just past Cheyenne, he jigged a little airplane dance around the flanks of

Laramie Peak and dropped in for our first look at the North Platte River.

There were just-greening hills on both sides of the Platte, the bigger ones to the west casting oblique shadows across the old furrows of the Mormon and Oregon Trails. They weren't the first thoroughfares that had followed the river's floodplain: for centuries the Pawnee Trail had also run along its banks. In its ancient path the poles for the earliest transcontinental telegraph were set, and their upright rows marked the route, first of the railways, then Interstate 80.

But the Platte — too thick to drink and too thin to plow in the settlers' homily — wasn't just a highway for humans. Since early in the Pleistocene, the Platte's waterway has been a defining feature of a considerable segment of North America's Central Flyway — the siliceous knolls along its banks having given their name to the river's most spectacular migrant, the sandhill crane.

To speak of this river's banks is to perpetrate something of an illusion, however, because for much of its course the Platte is only partially enclosed; mostly it's a gravel-paved hydraulic right-of-way scoured across the plains. Downstream in Nebraska its channel is not even a single right-of-way. Hundreds of yards apart, one road sign after another announces PLATTE RIVER as the interstate highway bridges four or five entirely separate large streams and a score of smaller creeks — all that remains of the river the Pawnee knew.

All that remains, because since the 1880s the Platte has been whittled away for the irrigation that in this land west of the hundredth meridian is the only way to grow the fields of hay and corn that fill its broad valley. Lately, the demand for urban water and hydroelectric power from as far south as Denver has shrunk the Platte even further, and less than three quarters of the river's original flow escapes the grip of forty-odd dams, which are only the vanguard of an additional score on their way from drawing board to earth-moving machinery.

Without banks, it was hard to tell where the Platte's boundaries were, and whether the river was a foot deep or ten. Probably not the latter, because trees — tough, gully-adapted hickories, cottonwoods, and elms, some with trunks thicker than fire hydrants — tangled up from its rocky bed, defining the Platte, from the air, as a wide woody hedge snaking between hay fields.

"Ought to just plow the thing and get it over with," George said testily. It was the end of a long day.

"Tell you what, Mister," he concluded. "Tomorrow we're gonna get up early and just keep flying bigger and bigger circles. Sooner or later, we'll intercept that old bird."

I was so tired the plan sounded almost reasonable. But the day before we lost her, Amelia had flown more than 450 miles. By now she could be anywhere from the Texas Panhandle to Hudson Bay — if she was still alive, was still carrying a working transmitter, and was still on our side of the Rocky Mountains. None of which we had any particular reason to believe, I bristled, since we were talking, for God's sake, about the whole western half of the continent.

George looked crestfallen and handed me the mike to call flight control at Douglas, Wyoming. The guy said to come on in. No traffic, only one runway. Vose had the little paved strip lined up, and was just off touchdown when he gasped and shoved the throttle forward. Four six nine bucked lower, then caught and pulled out. As we swept above the pavement, I saw two boys below, their bikes spun to a halt, waving madly at our undercarriage.

"Give me that!" George ordered, grabbing the mike. "You've got civilians on your runway, Mister. Kids. We almost clipped a couple."

"No, no, no," came the easygoing reply. It was a young fellow named Norm Reims. "You folks are down at the old, closed-down strip. New field's out by the river, three miles north of town."

Town, Vose and I found, was pretty small, and cruising through it in Norm's loaner car we found that the only place to stay was the Super 8 Courts, a pumice-stone approximation of eight Paiute hogans. Ordinarily I'd have thought the cabins were pretty neat, but in exhaustion we clumped without interest past their "Closed — Call Again" coffee shop and, with shots of George's gin and orange juice, washed down our wieners and crispy snacks and fell out on the beds.

The last I remember was Vose scowling at his maps. Then, out of the dimness where dreams begin came falcons: the big, sparkly eyed mortuary peregrines of the tribes. They were all female: priestesses somehow the same as the adolescent girls — indomitably potent in their screeching, after-school flocks — I'd both feared and been bewitched by as a boy. Now they were back, fluttering shadowy wings around my face, then fading into the darkness. There, time after time, I'd try to follow.

Finally, I couldn't stay asleep.

"It's morning," I told George. "You said early." I had the bathroom light on as Vose groped for his watch.

"Four a.m. is not early, Mister. It's not even morning."

It wasn't, either, for as we stumped out to the car the air itself was like lampblack. But it was also still: the norther had died. We parked behind the flight office, and while George left Norm a note thanking him for the use of his vehicle, I hauled our duffels out to the plane. It was too dark to see, so I climbed in and sat there, shivering, wondering what I was doing huddled on the decomposing seat of a bad-running old Cessna parked on an unlit airstrip in Converse County, Wyoming, nearly 2,000 miles from where I had last seen an illegally radio-tagged peregrine. Bob had been right, after all. The month of following telemetry falcons across the laguna, the close call with Ward, and our long trek up the Rockies had left Vose and me unable to give up our dream, though by now it was clear Amelia was really gone.

Then George climbed aboard and cranked up the Skyhawk, estimated the lay of the invisible runway, and pulled us away from the pavement before we'd gone far enough to angle off its edge. By 1,000 feet, Douglas's handful of lights had vanished into the blackness, and gradually everything changed. Wrapped in depthless velvet, we rose into a swarm of stars that dangled like shining ornaments around the plane. Suspended in the darkness, George and I also hung motionless, out of the space and time we inhabited on the ground, and all at once, come what may, I was glad we were up.

Vose reached over and thumped the receiver.

"You planning to turn that thing on?"

We'd been at this so long I didn't feel right without my stereo static, so if only for tradition's sake I punched in Amelia's #.759 and flipped the switch.

After a minute I reached down, waited for my hand to steady, and plugged George's headset into the other outlet.

It was too dark to see his face, but I watched him jiggle his earphones in puzzlement, then look over.

"Sounds pretty close . . . ," I began, but

Vose's whoop cut off the rest. Not far to the south, out of the silvery sky where dawn would be in an hour, Amelia's electric cheep came beating steadily up the prairie. Skimming the tips of the inky bluestem, she might have wondered what was going on overhead as the drone of a light plane throbbed round and round, wagging its wings for joy.

Back in Douglas, where George had insisted we go for a celebration breakfast, Sunday brunch meant an early-service crowd of Methodists at the Holiday Inn, and we got there just in time to grab a table. Next to us a barrel-bellied rancher in a snap-button dress shirt sat across from his wife, whose playing-field voice let Vose and me learn that she was both the music teacher and the track coach at the high school. Between them was an older lady, and all three stayed busy nodding to everybody in the place except George and me and a wizened little guy in a crumpled straw hat who sidled in as we were figuring up the tip. He might have been a broke-down ranch hand, but our hostess recognized him as Neil Allman.

"Your family ran the drugstore here, years ago!"

Neil pulled his longjohn top layer off a denim workshirt and said he'd come back to check on, maybe sell, the old building.

Our neighbors had shoved back their chairs and were on their way to join Vose and me at the register.

"Neil," said the hostess, "this is Vernon and Presly Watson and their neighbor, Ms. Willis."

Allman lit up. "Mrs. Willis! I remember you. You were my second-grade teacher."

"Neil," said Ms. Willis sternly. "I most certainly was not. That was my mother; I was two grades behind you, all the way through."

Outside, above paisley prints and summer Stetsons, bells chimed from Douglas's steeples; with the norther past, spring had come back to the plains. On both sides of the airport road, meadowlarks strode jerkily through the still-bare fields and yellow-headed blackbirds buzzed in the culvert tules. Vibrating the top strand of barbed wire were dickcissels, branded with a big black V across the males' yellowish breasts, calling out their "dik, dik, dik, dickcissel" names to their hidden mates. Above them black-faced bobolinks fluttered in circles, showering their masculine song claims over as much territory as possible, and next to the airfield a kestrel swayed its hook-beaked profile on the hangar power line. Reims had it in his binoculars.

"You guys missed the eagles," he moaned. "Five balds heading north past here at sunrise, right after the wind died."

George pointed a finger at me, confirming his theory that Amelia had followed his long-ago telemetry eagle's path, and proudly told Reims that our peregrine was also a canny old bird who knew when to use her energy and when not to fight the wind. I was afraid that by now her energy might have carried her out of radio range, but back in the air, at 2,000 feet I got a good signal, north by northwest, and Vose happily fell in behind it.

For now we knew Amelia's route was not the alpine course the biologists were sure would stop us. There was no need to enter those high passes: she had a better strategy. Lifted by the orographic currents pushed up by the cordillera's eastern wall, in the company of snow geese, sandhill cranes, and a hundred smaller species we could not see, Amelia was riding the mountains' uplifts north. Then, when headwinds had stilled the passage of the earth beneath her, she'd waited two whole days, hunting and feeding perhaps, and when the air she lived in would once more carry her toward home, she had again launched herself forward, fueled by the stolen energy of the prey-bird breasts tucked in her crop.

The mystery, of course, lay not in Amelia's choice of when to travel and when to rest. The enigma was what gave her the determination, day after day, to keep jumping into that oncoming airflow — to struggle with the rest of her fellow migrants, swimming against the oncoming current, up the spine of the continent — which at this point happened to be the Porcupine Creek badlands.

Maybe what propelled her was no more than the biotic pull of the light that lasted longer with every gained mile of latitude. That's what most biologists would say, yet it was known from recovered bands that even tiny white-crowned sparrows came back, all the way from across the continent, not just to approximate wintering and nesting regions, but to the very same bird feeders and identical stands of springtime nesting brush that they had occupied the year before. For months, those small creatures had unerringly remembered such specific sites, and I was sure that these days of less and less darkness must be stirring in Amelia equally precise, irresistible memories of her always-sunlit summer home. A place that would also hold other recollections. Wordless images of her mate: preening maybe, offering bits of prey, then ripping high cartwheels from the air before their ledge.

And what of all the others? Were not similar luminous likenesses of their home places guiding back every pintail and widgeon, grebe and plover and sandpiper? Every tree sparrow and flock of warblers that, below and beside Amelia, were simultaneously pressing their individual ways north?

Beneath my wing, a spray of white drifted across the prairie. Snow geese. Resting after an all-night haul, they were beneficiaries of the plains' new cereal monocultures. Four days ago they'd eaten the last of the fallen paddy rice left by Louisiana combines, and now with a winter's worth of yellow fat packed between their downy breasts and dark pectorals, they might feed only sporadically until they dropped onto the awakening tundra north of the sixty-sixth parallel.

A thousand feet up, midway between the

geese and '469 a translucent layer of big pale flakes slid by. Even without field glasses I could make out the long yellow beaks and black-bordered wings of pelicans: the snowy, inland-living relatives of Risebrough's Channel Island browns, moving across the landscape in a loose V formation. Slow flappers like pelicans, geese, and cranes travel in those trailing lines because each wing's downstroke creates a revolving disc of turbulent air called a vortex wake that, defined by high-speed photography, looks like a vertically oriented smoke ring. But they are rings of power: each one a firm, swirling bubble of wind that can buoy the adjacent wing of its neighbor, who then passes the gain on to its nearest downwind companion.

Vose had spent much of his life in the air, but hearing the flight strategies of other airborne creatures, he said, still gave him a better feel for what our old Amelia was doing. For now, aloft in its midst, George and I had become part of that ancient river of migration. Most of our companions were too small to see, but below us by the tens of thousands flew godwit and plover, sanderling and knot, wee kinglet and wood warbler — all flung forward, a hundred or a thousand miles a day, by the same flurry of hollow bone and fast-striking heart with which each bird worked its own small miracle of return.

14.

Airborne

By now the ground had warmed, and as its heat swelled the air Amelia's speed went up. But not steadily. Ridges of granite protruding from the buttes took in more of the sun than the adjacent grassland, buckling aloft mounds of rising air whose contours mirrored the corrugated topography that generated them. Stroking into those invisible currents, Amelia would ride their waves, her signal rising toward us, then dimming as she roller-coastered down the thermals' far slopes. Minutes later we'd bump upward on the same unseen hump of air.

Even with Amelia's downhill bursts I figured we'd overtake her in half an hour, and when I looked up from my numbers the badlands had given way to cattle country. Sundance, south of the Belle Fourche River, appeared below my window, and George folded the last Wyoming sectional and reached for Montana. Now the pastures held double snow fences, and in rocky meadows edged by fir and viridian spruce the Herefords that the snow breaks were there to shelter made do with less grass. Among the trees a thread of rough water snarled silently northward, roiling a pink emulsion of chert that had been bumped, abraded, and ultimately ground so finely during its tumultuous trip down from the peaks that, camped nearby with his partner Meriwether Lewis in 1806, William Clark wrote in his journal, "I observe great quantities of . . . stone thrown out of this red Stone river," a name later modified to the Powder.

I glassed the Powder's course for Amelia, but we were too high for a glimpse, and somewhere above its last rapids we swept past her. That meant we'd have to stop and wait.

Twenty miles downstream Vose tapped his window. "No town, but there's a sort of runway. It's short, with trees, but we're going in. Steep."

No kidding. We skimmed a bare-topped double hill that bordered the field, flew downwind, then swung around and, with the stall button squalling, dropped over a wall of spruce. Four six nine banged the dirt and bounced, but the runway ran so sharply uphill it brought us to a halt well before its upper end. Out my side was a sagging plank barn; out of George's was evergreen forest.

With the engine off and our doors open we could hear the soft mountain air combing the spruce and rustling the cottonwoods down the hill. In the distance a redtail shrilled its courtship whistle, and with Amelia's transmitter still coming in clearly, we climbed out onto last year's snow-flattened grass. A haze of golden pollen floated over the meadow, and for a while we hunkered under the wing.

"Is this kind of flying difficult? Like yesterday, fighting the wheel all afternoon?"

Vose shook his head.

"You'd think so. But no, my hands just do it." He turned them over, studying their heavy fingers. More than just slightly, they trembled. "Been very few pilots flying up into their eighties. I hope to be one of them."

Behind everything George had done lay that same steady aim. It was the old-line Mainer's self-determining vision. The long perspective, and he, like others, had done it right. Chosen a reliable tool: a standard, single-engine light aircraft, and like someone learning to use a complex lathe or harvest combine or diesel fishing boat,

Vose had made its technology his own. Made it, to him, a commonplace.

That's what you did in Maine, he said, during the thirties. Then you put that knowledge to use. Art came, if at all, with the years; intrigue, danger, and the conquering of fear had nothing to do with it.

At least it hadn't for a long time.

"I used to have dreams of falling," Vose admitted one night, after hours of fighting the mountains' turbulence, "way back when I was a beginning flier. I'd be wallowing around the sky, up on one wingtip or the other. Then in my sleep I figured it out."

Tilting back from our dinner table, he stretched his long legs.

"I'd stick my feet out through the bottom of the plane. Then, right before it hit" — Vose opened his arms, spread his palms, and smiled — "I'd swoop up with my wings. Set down soft as you please."

When George had left his Big Bend ranch for the Texas coast more than two months before, he had stopped building the house he'd had under construction for nineteen years, so I wasn't exactly throwing him off schedule. After shutting down his flight school and moving west, the first thing he had seen to, even before

259

hauling in an aluminum house trailer, was his landing strip. It was a strange landing field, because directly in the center of its touchdown zone stood a giant Thompson's yucca — an ancient desert plant that Vose couldn't bear to chop down. So he had to swing a rapid-fire dogleg around it every time we took off or landed — an imposing requirement that effectively deterred most fliers from unexpectedly dropping in.

So far, the walls of Vose's house were only chest high because to build them he had to haul water from a well 80 miles away, and only what was left after irrigating his sixteen Arizona ash saplings and six little cottonwoods went for making bricks.

"Not bricks. Mud-and-straw adobes," he corrected, indicating a size about that of a small suitcase. "Mix 'em myself, and when they're hard I coat them with polyurethane."

He grinned at the secret. "Last a hundred years."

We still had a little gas left in the cans behind the seats, and after I found a ladder in the old barn, George pulled off his coat and lugged the fuel up to dump into our wing-top tanks.

"Better to not have that stuff in the cabin with us," he said, "on this takeoff, anyway."

In the distance, beyond the Little Bighorn River we could see Wolf Mountain. Nimbo-cumulus were building behind it, but here in the foothills it was warm and still in the sun. Vose and I drank water, made our pit stops, and as usual munched up the tuna and crackers that were supposed to be our survival rations. Beside the Skyhawk, four black-billed magpies bounced on springy legs in and out of the grass, and I tossed a cracker in their direction. That only offended them, and dragging their floppy blue tails they leapt up into the firs, chirring resentment across the quiet meadow.

Ten minutes later Amelia slid over the radio horizon. We couldn't put off leaving any longer, and for once Vose was tense.

"This is a tough field. Uphill. Even with a climb-propeller I got to get everything perfect."

We slammed our doors and bumped down to the bottom of the landing strip. There George swung us around and let '469 roll back until its tail pressed into the edge-growth hemlocks. That was all the room there was.

Sitting halfway into the brush Vose ran the engine to 2,500. Both ignition magnetos working, still rough. Then he idled back, waited a second, and did it again. The third time we hit redline he stepped off the brakes and we leapt raggedly out onto the runway. As we jarred across the dirt, engine shaking on full throttle, beyond the field's uphill end I could see the pair of stony hills we had buzzed on the way in; by the time we were halfway there I could tell we weren't going to make it.

But we had already rotated; lifted our nose wheel 10 feet in the air, where it was too late to shut off. Then the airstrip vanished behind us and we were over rocks, barely, flying up the side of a small mountain. I could see the same pebbled detail as I would sitting on the ground, and I braced to hit.

Instead, just before impact we buoyed a little, cushioned on the last-ditch film of air sandwiched between our wings and the earth. Ground effect. Second after second that compressed, salvational layer of air held us just off the hillside as I gripped the door handle and looked at Vose, who was frantically checking for a way through the trees on either side. Ahead, between spruce still 20 feet higher than our wings, an opening flashed up and instantly George banked us 90 degrees left, hoping the gap was more than a few yards deep. It was, and we skimmed through the horizontal slot of a long mountain pasture, wings flanked by lofty conifers, above whose skyscraper tips we gradually rose into the clear.

It was 80 miles, all downhill, to Miles City, and if the plane hadn't been making such a racket I would have sat back and calmed down. At least Amelia was solidly on the receiver, invisible as usual but front and center, straight up her usual heading. As George eased the throttle and we sank over the last of the forest, we floated above a maze of quartzite bluffs, and sailed out over the Yellowstone River, which ran right down to the airport.

Miles City had a long, wide runway, but as we touched down on it '469's clatter racketed up so sharply that Vose, still jittery, pulled us back into the air. When we'd come around and settled again, the noise was even worse, and I was happy to see our propeller stop next to a row of repair hangars. Beneath the engine lay the problem. Scarred from bouncing down 200 yards of asphalt, the ruin of a cigar-box-shaped aluminum chamber lay between our wheels. Connected to the plane by a section of fat rubber tubing, it hung from the Skyhawk's forebelly like a disemboweled intestine.

Gingerly, I lifted the metal chamber. Busy with his flight log, George didn't look up. "Carburetor heat box. You can saw off the riser hose right at the cowling."

I couldn't bring myself to do it. That aluminum compartment was part of the plane — a complex, important part, surely included by Cessna for a reason. Weren't we going to need carburetor heat farther north?

Digging in his pocket for a knife, Vose saw he was going to have to perform the surgery himself. "Need that warmth more down south," he said, groping under '469's throat, where the flailing box had battered her belly pan. "Prevents icing up, mostly in humid air," he explained, whacking off the carb's rubber viscera and tossing its amputated length over my seatback, where it clanked to rest amid our gas cans, spare parts, and emergency bottles of gin.

As Vose went for fuel I tried to pick up some weather, for Amelia was already north of the airfield. That meant we weren't going to have much time in Miles

City, though I took hope from a Billings weather report that the clouds we'd seen building behind Wolf Mountain were the forerunners of an advancing cold front whose leading edge was now stalled over Fort Peck Lake. Amelia was traveling in that direction, and when she ran into its headwinds, I hoped they would force her down long enough for us to get some repair time.

Then George came back with the fuel truck. Without his parka he was shivering, but we figured it had to be somewhere in the plane, and not long after Amelia's signal once more vanished, we paid for our gas and, bearing the remains of our severed heat box, took off for Fort Peck.

The dark underside of the arctic front that had brought snow to the Canadian prairie was clear from 1,000 feet though it was still a long way off, so for a while we rode north on smooth air. By that time we'd climbed above the scarp that separates the Yellowstone watershed from that of the upper Missouri, which spreads out into the Fort Peck impoundment. Just as I'd hoped, Amelia had halted well ahead of the front, so I looked for a refuge where we too could go down when the wind hit. What I saw on the map was Hell Creek.

From my window its banks looked like the rest of eastern Montana: sandstone bluffs sprigged with bunchgrass and yucca. But on Hell Creek's eroded terraces, in 1902 Barnum Brown of the American Museum of Natural History discovered the first complete *Tyrannosaurus rex* skeleton. Since then, fewer than fifty individuals have been found worldwide, most of them exhumed right below us by Jack Horner, paleontology curator of the Museum of the Rockies in nearby Bozeman.

Horner's finds — along with those of tyrannosaur expert Tom Holtz, Robert Bakker, and their mentor, Yale University's John Ostrom — have gone far beyond carnivorous two-legged theropod dinosaurs, however. Their discoveries have helped to establish that birds — in the entirety of their kaleidoscopic color, intercontinental migration, and variation into more than 2,000 species — have all stemmed from a single small sprout on the tyrannosaur branch of the dinosaur family tree.

Like much of the Front Range foothills that Amelia had been leading us over, the badlands below had originally formed the western edge of a north–south ocean called the Western Interior Seaway, which had once filled the basin of the Great Plains.

During the first blush of Hell Creek pale-ontology, it was thought that the wading-bird, near-shore ecosystems had been the home of the closest ancestors of living birds, the ornithurines. But the discovery of *Apsaravis,* an 80-million-year-old pigeon-size bird that lived deep in the interior of Asia, showed that even during tyrannosaur times non-aquatic avian forms closely re-lated to modern birds had also inhabited the center of North America. By the time Alvarez's asteroid generated the worldwide cloud of debris and acid rain that delivered the final blow to a dwindling reptilian world, there were already hot-blooded, freshly wing-feathered avian lineages living alongside their *Velociraptor* relatives.

One of these, *Deinonychus* — the human-size predator made famous by *Jurassic Park* — revealed claws whose fossilized re-mains are almost indistinguishable from the talons of modern birds of prey, and it is thought that *Deinonychus* employed its great hind claws by rearing back on its rod-reinforced tail to free its legs to kick and slash. Exactly the way our trapped Padre peregrines and my Dad's gun-shot redtail used their stiffened tails to tip back-ward and bring their razor-tipped claws into play — behavior, Holtz maintains, "much more closely associated with modern birds than with reptiles."

Hell Creek's fossil beds were slipping behind '469's tail fins, but by Miocene times, according to the Black Hills Institute of Geology, its formations already contained a healthy representation of early birds: feathered creatures whose lines of descent had escaped the planet-wide Cretaceous extinction. Their world, newly vacated by their theropod forebears, contained loons, flamingos, cranes, waterfowl, ancestral shorebirds, and comparatively modern raptors — probably including falcons.

Amelia, too, had come back to her place of origin.

15.

Canada

Five miles back from the lip of the palisades ringing Fort Peck Lake was a little prairie landing field, and across the intervening valley I sighted a path toward it through the clouds. Vose started his standard broken-field run between the norther's forerunner squalls and had just banked right when we hit a monster current of wind that in an instant rocketed us upward, rattling pebbles of rain off our windshield.

"Virga," said George, leveling '469. Embedded in the storm, we swept helplessly toward the sky. "This rain's being kicked all the way to the stratosphere. It'll freeze up there and fall as hail."

I remembered the meteorological process. I just hadn't expected to ever be part of it. Yet, like the glider pilots I'd known, who were invariably seduced by the euphoria of lift, I had grown attached to the thrill of riding the daily mountain thermals along the Rockies' frontal wall. Death was

269

down, after all, and we were going up. Why not let that monumental surge take us as high as possible, and then, like a Capulin raven, soar right off the top when its boost petered out?

Surfing that big wave wouldn't have bothered my old stunter partner, but the long-distance voyager in him had to look ahead, test the air. He nosed us down and pushed in a little throttle. Then more, and still we went up. If we didn't have the torque to pull out of that freight-train rise even with the help of gravity, Vose said soberly, we were going to have trouble coping with what was sure to be coming up on its far side.

Two minutes later we sailed over the top of the ferris wheel and, nose still lowered, and felt the bottom fall away. It was what George had been afraid of, and as our lift vanished he shoved the throttle all the way in and tried to climb against the downdraft. In seconds we could see it was no use.

A knotted fist of wind pressed down, obliterating the pull of our tiny propeller. Feeling its iron grip, George set his jaw. Now there was only one option, and sliding the wheel forward, he dived. We were going to power down, even faster than the air currents, hoping to build enough momentum to rip free of their downdraft.

Below, the Missouri rotated up to fill our windscreen and I could feel the Skyhawk's airframe begin to quiver with unfamiliar speed. As the gray-brown earth rushed up to meet us, in a futile gesture of self-preservation I shoved my seat as far back as possible, thinking how men instinctively raise shielding hands in the face of gunfire.

Aimed at the ground, on full throttle we went down, second after second, building velocity. But there was a limit to how much airspeed the Skyhawk's airframe could handle, and suddenly its quiver exploded into a right-left shimmy that hid the hills, the windshield, even the dash in a blur of vibration. Both hands gripping the dancing yoke, speed beyond 150, Vose managed to lean over.

"We'll get out of it!" he yelled.

I shut my eyes and hoped so. Then the seat slapped up beneath me, and we had lift. Only for an instant, and again we dropped, but now we were on the edge, fighting to pull our dive-built impetus out into the horizontal thrust that could wrench us away from the down-shearing column of wind that had seized our lives.

Shuddering with too-high speed we inched away, stripping from '469's trembling fuselage the gradually diminishing gusts that finally left us floating free, suspended on fragile aluminum wings, 800 feet above the valley.

George started to climb. Then he thought better of it. It was time to get out of the air.

The only building at Jordan's airstrip was a phone booth, standing like a tiny, transparent temple surrounded by bare brown hills. Beyond, a single-family cemetery — ten headstones, six unoccupied plots — was the only other sign of human life. But I needed to put my feet on the ground, and with Amelia steady on the speaker and Vose rummaging for his coat, I slid through the barbed wire and went for a walk.

There was absolutely nothing to see. Behind the cold front it was cloudy and still, for spring had made it only as far north as Douglas. Somewhere around Miles City we had shifted back a season, and I crunched over starched winter grass empty of moisture, empty of insects, waiting for rain and birds and the coming sun. Dormant was all right, though: it was the earth, where I belonged.

Tawny as the prairie turf, three antelope watched from a ridge, ready to run. Farther on, a pair of cruising ravens bent down their heads for a look at my trudging form, and little by little, as life trickled back onto the plains, somehow it was done. The prairie itself — its minimalist grass and earth, lingering there on the cusp of spring — had erased the sickening vertigo, the falling terror of the hailstorm, nurturing me with its terrestrial succor.

I went back to the plane. Still in his shirtsleeves, Vose huddled on his seat.

"Know what happened to my coat? I left it on the wing, back at that mountain strip. Whump on takeoff was it hitting the tail."

Focused on our own imminent whump into the hill, I hadn't noticed, but I dug around for my extra windbreaker. Its yellow sleeves concluded 3 inches above Vose's watch and the only button he could snap was its collar.

"Know what else?" he said, looking strangled. "We've had our heads in the clouds, Mister. Today Amelia's flown nearly five hundred miles; if that cold front doesn't knock her down, in half an hour our old bird will be in Canada."

I hadn't thought of that. It had been too

much to expect we'd ever get to another country. But we still had a good signal, so what real difference did the border make? On back roads, people drove in and out of Canada all the time without going through Customs.

"Entry clearance," George said firmly. "Can't fly across an international border without clearance."

Canada didn't seem very international, but when it came to major aviation law Vose was as straight an arrow as they came, and as I watched he got on the radio to the Air Control District Office in Regina, 100 miles east. It was the nearest port of entry, but they were about to close for the day.

"Come on in, then. Here: nowhere else," pleasant but adamant Officer Kristy Fishell said. "Early in the morning, we'll fix ya right up."

George said she didn't understand. By then Amelia would be 200 miles away — so far we'd have no chance of finding her — but his static-garbled explanation was lost in transmission, and until tomorrow our prohibition remained in force.

What we needed was somebody able to both grasp what Vose and I were up against and sufficiently highly placed to override conventional protocol. Get us permission to register, sign in, whatever the hell they wanted. But to do it later.

Reaching people at that level was one of Binder's specialties, and with misgivings, after having last seen him hurrying across the runway at Abilene, I called him at the office during our next gas stop. My initial penance was hearing how there had been no commercial flights out of Abilene and that he'd been forced to wait all day to get an evening bus to Dallas, then spend the night before catching a morning plane back to Austin.

Finally Bob relented, graciously turned away from his legal briefs, and went to work on our behalf. He said he would try to reach a colleague in aviation law while I held on. Down the hill from my little phone booth I could see the town of Glasgow, straddling the Milk River; 25 miles north, its tributaries flowed out of Canada, and my hopes were high as Binder came back on the line. I was right about his connections: Bob's associate had a buddy at the Federal Aviation Agency who said it was fine with him for us to go into Canada on an important scientific quest. But of course he couldn't speak for the Canadians. And he didn't know any authorities up there well enough to call them late in the day, at home.

Also a pilot, Binder said he'd try Canadian Air Control himself. That took a long time, during which George came over to say that Amelia's signal had gotten real faint. Still on hold, I hunched my back against a blustery wind, looked at the empty hills that stretched away on all sides, and felt my confidence vanish. After all that Vose and I had been through, what if we could go no further?

Then Bob was back, with bad news. The only port-of-entry office he could reach was 350 miles away on the Pacific Coast, where it was an hour earlier. If we could get there in forty-five minutes we could go into Canada via Vancouver.

I took a breath: we were done for. There was nothing else to do, so I headed back toward '469, doors open but otherwise ready to go, out on the runway. George was fidgeting, checking the engine and listening to the once-more vacant buzz of static on our scanner.

It just wasn't going to end here, I decided, and grabbing the door handle I swung into my right-hand seat.

"Request's been OK'd," I declared. "We're clear to cross the border."

Stunned, Vose shook his head.

"Alan," he said with admiration, "I wish I had your way with words."

On our second day in Canada, we saw cranes. It was dead calm on the runway at Lloydminster, and as George sipped his coffee and listened on the receiver to Amelia, who had roosted only 6 miles from the airstrip, I pulled '469's wheel chocks. In the stillness, the clank of our tie-down chains sounded like an ocean liner being unmoored, but as I stood up I'd already heard the great birds. From the dawning sky a few faint calls had come floating down, and though their makers were far out of sight overhead, I was sure they were the voices of sandhills.

Birds of Heaven, Peter Matthiessen calls cranes, and like no other sound, their gurgling, ethereal cries always brought life to the day. For cranes were the birds of my youth. Harbingers of my seasons, they were the creatures who from a world beyond my own drew in the sadness of autumn, an ancient solemnity that for me transcended the round of sports, Halloween, and Christmas that meant so much to everyone else.

Instead, I'd yearned to live out there on the vital, scary edge of the lives they led.

Lives larger, older, more vital than those of the people I knew, and during years when the northers held off past when the leaves had changed, and the sandhills were slow in coming I worried. But then, on some ordinary sunny day, I'd hear what no one else heard — a faint, musical bugle drifting down 2,000 feet. Looking up, I'd pick out ten or twelve gray specks dotted against the blue and feel my heart clench, and soar, and make me yearn not just to go with the cranes, but to *be* one.

In the room where I slept when I was four or five hung Antoine de Saint-Exupéry's drawing of the Little Prince being carried aloft by "a migration of wild birds." I remember my fascination with the prince's effortless escape, but it was more than forty years before I realized I was not the only one who held the same longing to go up, and that all over the world the desire to be carried aloft by birds had centered on cranes. Mostly that was because big gruids seem almost large enough to carry a human and can climb so high that, like peregrines, they disappear from earthly sight. Only their ethereal voices waft down, trailing from heights that, until the advent of aircraft, remained hidden in the mists of human imagination.

Among Native American groups, cranes ferried the souls of fallen warriors to the Great Beyond. Among the Cree, Crane carried Rabbit to the moon, and Aztecs styled themselves as People of the Crane. The circled Hopi drawing of a sandhill's footprint has become today's peace sign, and in Moscow, the aerial ballet called the Flying Cranes is an international sensation.

But that spring morning, from Lloydminster's quiet runway I scanned every inch of sky with no sight of the magical birds. Finally, it was George's old fighter-pilot eyes that picked them out: through my binoculars, a scraggly V of twelve, headed northwest.

"Same direction as us," Vose waved. "Maybe we can pick 'em up in the air."

By the time we'd gotten clearance from flight control, taxied out, and lifted off the cranes were gone. It took a while to climb out, but then there they were, far above us, pale gray bellies luminous in the early sun.

George raised an eyebrow. "Better'n 3,000 feet."

I told him that by comparison this bunch was hardly trying. Sandhills have been seen almost level with the 18,000-foot peak of Denali, North America's highest moun-

tain, and on their annual migration between Russia and southern Asia, big white
Siberian cranes go even higher as they
cross the Himalayas.

George gazed up. "Figure this flock's
going all the way to Alaska?" He shot me a
look. "Like us?"

I told him it was possible. One of the
sandhills' primary breeding grounds is the
Yukon/Kuskokwim River delta. Others
nest on central Alaska's muskeg flats, while
a more ambitious group, like tundra
peregrines, surmounts the Brooks Range
to stalk and leap in courtship dances on
the lake-pocked marshes of the Arctic
Slope.

There was an even more distant possibility. Fifty thousand lesser sandhills from
as far away as northern Mexico fly on
across the Bering Strait at airliner altitudes, then continue more than 1,000
miles into Siberia. There, these little
brown "cranes from the East," as they're
known to the local Yakut, nest on the
deltas of the Kolyma and Indigirka Rivers.

Vose took all that in stride, pulling the
Skyhawk's nose up just shy of stalling in
order to overtake the cranes as slowly as
possible. In seconds we had flashed by, but
as we did I could see that these birds

weren't headed for Siberia. Only lesser sandhills go that far, and these were the big guys. Greater sandhills. With their outstretched necks, bulbous heads, and wings that spanned nearly $6^1/_2$ feet, they were so substantial it was like finding skydivers spread-eagled in the air.

As '469 pulled away, looking back through the glasses I could momentarily see the strength with which each crane drove itself forward on the flexible outer two-thirds of its cinder-colored primary feathers, each bird's flight helped by the updraft of its companions' wingbeats, for atop each bird's wings the tertial feathers — the crane family's uniquely structured plumage that on the ground fluffs out into its characteristic high-rumped bustle — bowed upward, creating the same convex surface as the Skyhawk's wing. That bulge, like the Cessna's bowed-up flight surfaces, forced the air sliding over their wings to travel farther than the straighter flow moving by beneath, creating lift.

In seconds the cranes were out of sight, but I wondered how, from the immense, water-pocketed tapestry sliding by three-quarters of a mile below, they would recognize their home territory. It was late in the year for cranes to be on migration anyway, and wherever this flock was going, it must have still been a long way off or it would not have climbed so high.

That was as far as I got with them, because '469's engine had begun to sputter. Switching rapidly from our right to left fuel tanks and back, George richened the mixture and primed the engine with the throttle. The old Lycoming motor's plugs might have fouled, he muttered, and as the tachometer needle wavered lower I grabbed my seat and the door-frame handle. Vose looked over the surrounding terrain, weighing the factors. In five seconds he had made his decision.

"Have to go back," he said calmly. "We got time."

In near silence we cranked around to the left, and thankfully I could see Lloydminster still on the horizon. Coasting much of the way, George tried the engine every few seconds; then a mile off the end of the runway he dropped our nose to pick up landing speed and, with me clutched tight to my seat, sailed in for a slightly crossways, bumpy, but nearly perfect touchdown.

As we rolled down the empty pavement I was angry, more so when Vose announced that this was not to be categorized as a forced landing, only a precautionary one, and when we stopped, without a word I went out my door. I walked a little way to the grass and sat down, legs trembling. In the distance, on what seemed like a far-off drive-in-movie screen, I saw a guy in overalls come out of a portable building and start for the plane. At George's wave that we were all right, he stopped, turned around, and then came bumping over the grass in an old pickup.

Still in a daze, I watched Vose and the kid — sixteen or seventeen — open the Cessna's engine bay. For a while they probed around, long enough for an even younger Canadian, maybe the first one's little brother, to join them.

I had just gotten enough strength in my knees to wobble over when the first young mechanic backed out of the compartment's gasoline-soaked interior, holding, on the end of its fuel line and throttle cable, '469's single carburetor. George and the two kids grinned at each other. Then they turned to me.

"Carburetter, eh," the older brother said cheerily. "Aboot came off!"

16.

Farewell

The next morning, either Vose's headset or his hearing was giving him trouble because he kept jiggling his earmuffs.

"Amelia. You got her yet?"

I nodded. Our Amelia was still there, still on her 310-degree heading, and at almost 54 degrees north latitude, we followed her for all of a day so long it lasted until an hour before midnight. George wondered how she could fly so far without stopping to hunt and feed, which she had done with such regularity back on the Gulf Coast. It was a good question. The high metabolism that gives peregrines the freedom to travel the globe carries a high price: they are always eating, their whole beings dependent on the flawless union of swift wing, gunsight eye, and clutching talon, and more than once, out on the flats, I'd seen a falcon swoop up from Padre's sand to snatch a tired songbird coming in from the sea. Then, without missing a wingbeat, with its dexterous yellow toes it would swing the passerine's lifeless body up to its beak, where it would disappear in two or three swallows.

Below us, I told Vose, the daily stream of northern migrants was still so thick I suspected Amelia was regularly snacking on the wing, and that she always went down at night near water in order to drink, and maybe bathe. Sure enough, in the morning she was late getting started, but by noon we saw the last of the grass. All the way up the Great Plains there had usually been, somewhere in our view, a farm with at least a portable shack and a double-track road. Now dense conifers stretched as far as we could see, and the only sign of humankind was our pale wing strut edging across unbroken boreal forest.

According to George's sectional chart, the last small settlement on our course was coming up, and since we'd whittled away our onboard stash of candy and pop-top mini-meals — in the cold, I'd taken to opening both ends of the can and pushing out a frozen spaghetti or bean-dip popsicle — that was where we had to go for provisions.

"Doesn't do to keep eating our emergency grub," said Vose as we climbed the steps of an establishment whose sign said tribal council store.

In five minutes we'd nearly bought the rough-sheathed plank building out. Paper shopping bags still only half full, George and I looked at each other with misgivings. On a final lap of the nearly empty shelves I came up with nothing but the store's last two cans of tuna and three huge containers of something labeled ersatz chicken.

Vose went further. At the cash box he plopped down his final item, a cardboard-packaged six-pack of Little Friskies Kitten Morsels.

I looked up at him.

"Emergency rations," he said, smacking his lips.

Though both of the Skyhawk's tanks were nearly full, an hour after takeoff George and I could see that before the day was over we were going to be in trouble. Here and there we'd spotted bulldozed landing fields — bush pilots used them all the time, and we'd been counting on their fuel caches — but the strips weren't marked on our charts, and being tied to Amelia's signal gave us no time to search for occupied or stocked airfields.

Two more hours passed before I saw a

narrow landing strip cut into the timber. For a moment we dipped lower, then Vose had the yoke back and was climbing out.

"Too rough, and no gas because there's not a shed. No place to tap a drum."

The last marked field I could find on our sectional chart was 20 miles east of Amelia's course, but there was no choice except to go for a look. The strip was grass, a rolling natural meadow, and off in the trees was a cabin with a big shed. Vose and I exchanged hopeful looks. To stay in touch with our girl we couldn't pass up any chance for fuel.

On the ground it was colder than it had seemed from the air, and snowflakes mixed with drizzle drifted across the grass. In the distance the pale green, shingle-sided cabin looked as though no one had been there for years, so I started for the shed.

"Maybe they've got generator gas. We'll leave money."

I was nearly there when I heard a small scream. One that could not have come from George. I could see Vose standing awkwardly on the cabin's little front stoop, peering into the inch-wide door opening. He had his head bowed, and even from a distance I could tell he was apologizing.

I decided to stay put. Little by little the

door moved inward, until I could see a frightened woman's face. Vose, who has possibly the kindest visage since Santa Claus, kept talking, gesturing back at the Cessna, then up at the sky. Finally, palms down, he wagged his arms like a falcon.

Slowly, the lady of the cabin set down the shotgun she'd been clutching, and by the time I got there she had come out onto her porch. Her trapper husband had gone in, as they say of anyone who's made the long trip to the nearest town. There was chainsaw mix and kerosene in the storage building, she told us, but it was pretty old.

I was leery of pouring that stuff into '469, but George told us that his mechanic had lowered the engine's compression enough for it to burn the bad Mexican gas he sometimes had to use out on the ranch. Even this marginal fuel, he said, ought to work.

I calculated that we still had just enough gas left to spike the chainsaw-and-kerosene mix up to combustibility, and '469 started right up. We were both strapped in as Vose ran a final check on the magnetos, when suddenly he cut the throttle and pointed out my window. There, blown by prop-wash but determined to reach my door, was our benefactor. I opened the latch —

to say good-bye, I supposed — but I didn't have a chance because into my hands she thrust a slab of homemade blackberry pie.

All that afternoon, as our fuel dropped lower, I watched the forest change. Every 50 miles the big trees were thinner and shorter, interspersed with larger stretches of frozen muskeg and stubby willow and alder taiga. But the trees were just the most visible part of what was going on below. As we flew, the land's richness in ecological niches, in speciation, and in biodiversity was declining with every northward mile. Along the Texas coast more than 500 bird species had lived beneath Amelia's wings; more than half that number bred on the northern Great Plains, but here at the far edge of the continent's boreal forest there were only about 120. If Amelia went on to the Arctic Slope, her prey base would be fewer than 50 species, though the absolute numbers of birds would be astronomical.

It was not just that larder of prey she and the other tundra peregrines sought in the Far North, though. There they would have outflown the only creature to regularly prey on them.

All the way up the plains I'd worried

about the great horned owls that hunted
Amelia's roost sites. Striking by surprise
with their heavy, feathered feet, the big
owls could be instantly lethal to a sleeping
peregrine. No predator kills as many of the
adolescent falcons reintroduced by conservation groups as great horned owls, and I
knew several falconers who had found a
cherished hunting peregrine as only a pile
of feathers after a great horned had discovered it during a single night of escape. I'd
seen the same thing myself. Co-directors
of Austin's Raptor Preservation Fund,
Shaun Ogburn and I had walked into the
hawk barn one morning to find six or
seven of our convalescing red-tailed and
red-shouldered hawks lying dead, their remains strewn below a great horned owl,
still perched in the rafters, that had forced
its way in through the wire-mesh windows.

Great horneds live everywhere in North
America except the Arctic, but below us, I
told Vose, was the home of the continent's
most stunning race — the silvery taiga
owls that, during the nightless weeks of
spring and summer, ghost soundlessly
through the thin stands of black spruce
that sprig up out of still-frozen marsh.
Here, more than ever, Amelia's salvation
lay in her tireless wings, and — except for

the problems we were having with fuel — I was happy she was now aloft eighteen hours of every twenty-four.

By evening, George and I had come to the last big conifers. They were the outermost trees: trees that had gone as far toward the pole as plants their size could go. They were the northern boundary of the vast boreal forest that stretches from British Columbia to Newfoundland — the rampart where tall forest ends. Beyond lay nothing but tundra.

Amelia had brought us to another frontier, for ahead as far as I could see rolled an olive green, billiard table–smooth plain. But there was no grass. The Finns call this circumpolar land *tunturi* — high, treeless ground — and below us its emptiness was broken only by dark lines of dwarf willow that scribed the tributaries flowing into the upper Mackenzie River.

George leaned my wing down.

"Leads right to the Arctic Ocean," he said. Then, after a pause, "Mister, we're here."

Amelia seemed to know it, too. For the first time in days she swerved onto a different heading, west of the river's wide channel. Gone was the careful, focused hawk who'd played the Great Plains winds, then ridden the Rockies' updrafts north. Now she was dropping back into the terrain of her heart.

But Amelia's turn had put us right on top of her, which meant we needed to set the plane down and listen for a while. Looking for a marked homestead, I spread the aerial charts across my lap.

George shook his head. "Can't land on this stuff, even if somebody had gas."

He went on to say that under its low bushes, the tundra was a bog. A green swamp of melted ice.

"Looks all right from here. But I've gone in after planes that tried it." He sunk the edge of his palm into his spongy seat. "Most never flew again."

So in big figure Ss we slowed enough to follow Amelia northwest; on the receiver her signal pulse was a bright, cheerful cheep.

"Sometimes," said Vose, "I get to thinking it's really her making those peeps. Calling to us."

For a while we flew on, neither of us willing to say what we both knew. Finally George tapped our two gas gauges, one after the other. Both were less than half.

"There's a couple of places back the

other way we can still get to for fuel. North of here, it's float-plane country."

With that, Vose pulled his steering yoke back and began to climb. Far below, Amelia had picked up even more speed, and I imagined her fierce face, only a few feet above the tundra, searching the horizon for the riverside cliffs her long wings were pounding toward.

In the perpetual sub-arctic twilight she might fly all night. Tomorrow she could be home. Perhaps in the morning she'd settle onto a big riparian bluff. Maybe along a tributary of the Mackenzie or, beyond the Brooks Range and another day's flight away, somewhere inland from the polar sea. There, she might come down at last in a land where even the taiga owls never venture. Might compose her wings on the ledge of an age-old aerie, notched into a cliff above that fabled river of the tundra falcons, the Colville — ancestral home to her kind.

Along its steep clay bluffs, because male peregrines tend to arrive home first in order to defend their ledges, Amelia's mate might be waiting; if so, within days there would be eggs to brood and protect.

Knowing what was to come, at 9,500 feet Vose leveled off and swept us into a broad listening arc. From up here Amelia's signal would be clear for almost an hour. But I shook my head. George raised his brow, and then, when I pointed over my shoulder, he nodded and slowly turned southwest, back toward the fuel pumps at Fort Nelson.

Still above Amelia's tundra home, neither Vose nor I spoke as the miles piled up, because to both of us she was still our gal — a creature we had come to understand, one whose choices we could sometimes anticipate, and whose aerial world of storm and wind we had entered on wings that followed only a little way behind her own.

For the most part, we had known Amelia only as a radio signal, an almost theoretical being. But for more than 2,000 miles she had been our guiding angel, and nothing less than the power of the dream carried in her heart and the fierce strength of her wings could have brought us all this way.

Soon the swiftly winnowing speck Vose and I had hardly ever seen would be all we would ever know of her. Yet she'd not been just an abstraction. George and I had flown where she had flown, seen the land that she had raked all day with her binocular eyes. And we had felt through our own fragile flight surfaces the same air currents, peered into the same mist and storm and rain that Amelia had known in every nerve, hollow bone, and airy feather of her hard-muscled body.

And that was best. To have joined her more closely would have been like conditioning a captured peregrine to the falconer's glove. Even worse would have been to make a pet of her. I'd seen captive-born fledgling peregrines hopping across their owners' sofas, flopping belly-down on the pillows. But Amelia had always been her own, untamed self, living and chancing death by the whims of luck and individual skill that determine the lives of all wild things, and I was glad we had shared her existence only from afar.

17.

Mounties

Parting with Amelia was such a wrench I'd
almost forgotten our legal dogleg back at the
Canadian border — forgotten it until, on the
way into town from the airstrip at Swift Current, our second refueling stop, a red
light–topped pickup cruised by. Inside were
two law-enforcement officers, and we were
going slowly enough for me to make out a
big green Royal Canadian Mounted Police
emblem on the truck's door.

That took a minute to sink in. Where
were these guys' crimson coats? Their
stovepipe boots? Whatever sort of Mountie
they were, I figured Vose and I ought to be
able to slip unnoticed out of town first
thing in the morning, and after we got a
cabin I flopped on one of the bunks while
George went in search of groceries. Mull-
ing over our non-official entry, I'd almost
decided to let Vose in on my subterfuge
when I heard his footsteps coming up the
stairs.

"There's Mounties all over," I called through our front screen. "Maybe we ought to fly out now."

I got up and, as the steps clumped across our wooden porch, went to let Vose in.

But it wasn't George. At the door stood a tall man with side-striped slacks and a crisp uniform shirt. On its pocket a small badge identified him as:

MARCEL FERLAND
ROYAL CANADIAN MOUNTED POLICE

Officer Ferland was stiff but courteous, and after waiting for Vose to get back, he said we couldn't talk there. It was a matter for headquarters.

On the way over, George leaned toward me. "No entrance papers," he hissed. "Could be trouble."

At the Regional Office, Ferland's staff of five gave us the once-over, though they didn't say a word about aviation regulations. All their questions concerned taxidermied trophy-game heads. Neither Vose nor I had the slightest idea what they were talking about.

Then the topic shifted to poaching.

Surely we had seen — or spoken over the radio to — other light aircraft, at low altitude. . . .

George and I were clueless. Vose even tried to tell them our Amelia-tracking story, but Ferland cut him off and, with French-Canadian formality, made a little speech. He said that his office had no facility for holding detainees, and that until it was possible to arrange transfers, some individuals — such as us, perhaps — might possibly be held in what could be thought of as a culvert.

Ferland paused while his men stifled snickers.

"What're these goobers gettin' at?" said Vose in a voice nobody could miss.

A concrete container, Ferland continued, that the Mounties sometimes used as a trap and holding pen for their occasionally garbage-eating bears.

Blankly, George and I stared at him. We had thought we were about to be charged with a violation of Canadian airspace. Instead, because we wouldn't admit to poaching taxidermied trophy heads we were to undergo bear-dung torture.

Vose shrugged. "Filthy as we are, shouldn't make much difference."

But nowadays the Mounties have computers, and while we waited — in a spotlessly clean office — our IDs were run through the system.

"Aha," Ferland's secretary actually said when she got the report. "The bear trap was a joke, but we knew you were up to something."

This time Vose and I confessed. Our fine was $90 (Canadian) for every day we had been in the country illegally. Then, since we seemed to be exonerated as poachers and game-head black marketeers, the squad sent out for lunch, including George's and mine.

While we were eating, Vose told about Amelia and how we'd lost her up in floatplane country. Then it was time to pay. As we began to dig out crumpled emergency tens and twenties it became clear how far down on our luck Vose and I really were, and gradually the smirks of the staff changed to expressions of genuine concern.

As we made our $720 fine Ferland dumped the money in his desk, then guided me out the door and down the front steps, with George trailing. At the cars he slipped a couple of Canadian bills into my pocket.

"From the officers," he said quietly. "It is not much."

Back in Miles City, Montana, Vose and I got gas, poured in a quart and a half of 50-weight, and checked '469's flight surfaces. Then, because from Canada I'd called Jennifer — asked her to come join us for the trip home — we headed over to the passenger terminal.

As we waited under a canopied walkway, a little commuter plane touched down and taxied in. Its stairway door dropped to the pavement, and there in the opening stood Jen. She had really come.

"You know why I flew all the way up here?" she asked after Vose had gone off to the bathroom.

"Because I called. Begged you to come." I thought for a second. "Because we need you to help fly the plane back."

Jennifer looked disappointed. "Remember what else you told me?"

"Of course. But tell me again."

"About the falcons. When you called, you said you were done with them."

I nodded, and because I did so, the flight home was a happy time for all of us. Just as George had said, Jen had a genuine feel for the controls, and with not much help she flew us most of the way to Texas.

It was a different kind of aviation. I had

the whole backseat to myself, though it was bumpier there, and a lot of the time I was on the verge of airsickness. But without having to hang on to Amelia's will-o'-the-wisp signal there was time for weather reports and even the chance to pay attention to them. In the security of a flight shack we waited out a storm Amelia would have dragged us through, and with a genuine airport-to-airport agenda we could plan how far it would be to the next gas stop and where we'd spend the night. Dinner — savory coffee-shop chicken-fried steak and mashed potatoes — was once more something to look forward to, and not once did I put so much as a dime in a flight-office food machine. But neither George nor I was really through. Not with peregrines; maybe not even with Amelia. The day before we got back was Vose's sixty-eighth birthday, and I had his present — *Airports of Mexico and Central America* — ready. Inside I'd written "Peregrines fly south, too."

It was a dream Jennifer could not share, but the issue didn't come up, because, starting in July, I was booked on the nature tours I led, giving slide presentations on a small ship winding up the Inland Passage to Anchorage. My first group was a note-taking gang of high school biology teachers, and we saw most of what they'd come for: sea otters slipping through the looking-glass surface of Glacier Bay, clusters of harbor seals hauled out on floating floes, and, in the distance, mountain-size bergs calving off the ice cliffs. Best of all were the cetaceans: black and white Dall's porpoises that looked like miniature orcas, huge fin whales, and, almost every day, blowing, breaching humpbacks.

Down on the water, from the Zodiacs we sometimes saw enormous burbling swirls that my teachers were delighted to learn meant that, far below, humpbacks were herding herring with rising screens of exhaled air bubbles. And every time somebody spotted a peregrine — always of the dark-breasted Peale's race that lived along that coast — I hauled out my scanner in the vain hope that Amelia might have missed her target on the polar tundra. Along the shore were many occupied aeries, some on the rocky coastal cliffs and others back in the trees, in the old stick nests of ravens and ospreys. From those nests the parent peregrines sometimes came out, chattering their territorial anger at us; yet to my surprise we also spotted silent, solitary pairs high overhead, apparently with no young to defend even during the height of breeding season.

Then my clients, as they always did, went home. After the next group had also come and gone I was tired, for in the field, complex problems often arose from the hidden currents of peoples' lives. But who was I, barely hanging on to my perpetually tentative relationship with Jen, to tackle such things?

The calm of Denali National Park's backcountry seemed like the answer. There was a lottery for camping sites, and I drew a good one, at the far end of the 80-mile-long road that winds through the mountains to the refuge's most distant boundary. The road used to be open to the public, and still is for the miners working gold claims beyond the sanctuary's southern perimeter, but ordinary visitors have to board a yellow school bus to reach their designated campsites, which we were not allowed access to until we had attended a lecture about our upcoming wilderness experience. Most of the talk dealt with etiquette toward bears.

"Remember," our earnest young ranger emphasized, "stay on the trails and you'll be all right."

"But aren't they *bear trails?*" I asked.

He held up the little pink permit — our passport into the far reaches of the park. "If you don't feel you are able to comply. . . ."

I shut up and, a complier like the rest, got on the bus.

The ride in was pretty neat, though. We stopped for moose and a silver-black red fox, and to peer at far-off grazing grizzlies, their humped backs just visible above the sedge. The next morning was even better: beyond my tent — pitched as far from the designated campsite where the bus let us off as I could walk before dark — good light spread over the tundra not long after 4 a.m., and as I peeked through my zippered door the movement roused a pale-ruffed caribou bull. Grazing just steps away, he bowed his neck and snorted, undecided whether to run or exercise his bravado and bluff me — a creeping thing sticking out its head like a marmot — off his turf.

Prancing back and forth, joints clicking in the peculiar fashion of caribou, he pawed the lupines, inadvertently flushing a calf, maybe his own, that was hidden in the heather and that scrambled away with a loose-jointed flutter of legs. It was the same scene, I thought, that for 100,000 years had taken place every summer day in these rich sub-arctic valleys — terrain whose herds of big game had drawn the first Asians across a temporarily oceanless seabed from Siberia.

As I wormed out of the tent I heard a small "tik ti-tikk" raptor scold. It came from a pair of dark-color-phase merlins hidden in the stand of black spruce behind my meadow. In the shadows, both appeared to be uniformly steel-gray, but when they swerved, quick as swallows, out into the sun I saw the dark beige of their breasts, covered with a scalloped brown inlay that ran down their bellies and onto their tails.

On one of those swerves the male flicked down onto the carpet of needles under the evergreens and snatched up something small. Merlins are generally bird-eaters, but on his way back to a snag that topped the tallest spruce a little tail dangled from his feet. He'd snared a vole. Perched, the merlin stared down at his lemony toes, which gripped the rodent's squirming body. Exactly like a peregrine with a captured duck, he bent to seize the vole's plump neck, twisting his tiny tomial teeth to sever its fur-covered spine; then, before bobbing his head to feed, he fixed me with a long, beady stare.

At four o'clock the Denali campers' bus pulled up to the roadhead I had walked away from the previous afternoon. Its new allotment of hikers poured out and nailed down rights to their 10' x 12' gridded tent sites, then assembled themselves into a walk led by a khaki-clad ranger. Sooner or later my orange-trimmed tent, pitched on a rise across the valley, would attract every binocular, so I sat on a rock and waited. Sure enough, before she'd even come to a halt, the ranger wanted to know if I had any idea whose tent that was. I told her I did and that, OK, I'd leave.

Then, under the field-glass scrutiny of her vigilance-keepers I broke down my camp and started downhill toward the bus. For all its unfettered wildness, Denali was a sort of Pleistocene theme park, not where I was supposed to be in Alaska, and suddenly it all came clear.

Not heading straight home was going to mean trouble with Jennifer, but Amelia was still in the Arctic. That, of course, could mean anywhere from Barrow to Ellesmere Island, but since she'd left Texas, Amelia had stuck close to the same heading — a north-by-northwest course that put her right on track for the one part of the North Slope where I knew there was a large breeding population of tundra peregrines.

I had been there before.

18.

Plane Crash

Under the sawtooth peaks of the Brooks Range, almost exactly a year earlier my Alaska Fish and Game Cessna 180 had dropped toward the handful of rooftops that was Bettles. As its pilot rolled the bush plane to a stop on the village's gravel runway, I thought how unexpected my arrival here was: three days previously I'd picked up my phone in Texas to find a message from Ken Riddle, calling from Fairbanks, where he and Rebecca, his bride of two months, were about to go into the Arctic on a peregrine survey. One of the Cancer Center's chartered aircraft had cracked up trying to get through a pass onto the Arctic Slope, breaking the back of one of Riddle's researchers; if I could make it to Alaska in twenty-four hours, Ken was offering me his slot.

Getting to Fairbanks was easy, but a late-season blizzard had turned back our pilot, who dropped Tom Cade, the former director of the Peregrine Fund, and me off just south of the Brooks Range, the long east–west mountain chain that separates most of Alaska from the North Slope. Tom had spent an enormous amount of time there; it was along the Colville River that he'd gathered nestlings to form part of his original breeding stock for the reintroduction of peregrines to the eastern United States.

The Colville was where, on a riverside bluff, he found Cadey.

Until the early 1970s, almost no one had bred peregrines in captivity, partly because caged falcons were unable to perform the towering aerial acrobatics of pair-bonding that wild-reared falcon chicks, having themselves lived athletic lives on the wing, also expected from their mates. Peregrines who grew up without having seen those displays, it was theorized, might form sexual unions anyway, yet for them to be viable breeders they needed to come from chemically uncontaminated stock, and one of the last places it was possible to find such a population was the remote nest aeries along the Colville.

Brought as a youngster to Cornell, as an adult Cadey was paired with a tiercel called Heyoka, and in the university's hawk barn, during the early seventies she produced fertile eggs. Her descendants — trained to live again on their own — ultimately propagated the first new generation of wild-born peregrines on the eastern seaboard.

All that seemed long ago and far away, however, as Fish and Game's blue-and-white Cessna, now without Cade and me and our gear, disappeared into the blowing snow, which was so dense that we could barely read the plywood sign, next to the airstrip:

BETTLES, ALASKA

64 DEGREES–54 MINUTES N
151 DEGREES–31 MINUTES W

POPULATION 51
ELEVATION 643'
LOWEST RECORDED TEMP. -70 DEGREES
HIGHEST RECORDED SNOWFALL 116.6 IN.
AVERAGE MEAN TEMP. 21 DEGREES

Back near the evergreens was a log cabin, 10 feet on a side, roofed with sheets of bark that had grown a 6-inch coating of moss. The cabin was set 12 feet off the ground on the trunks of four lodgepole pines because it was a supply cache, built to keep food safe from bears. It wasn't clear why the same defenses weren't necessary for the low-slung building next door, which was Bettles Lodge, except that it was full of people. Inside, a pair of French-Canadian prospectors, stubby, wildly bearded men, glared at a group of backpackers' Gore-Tex parkas and multi-pocketed wilderness shorts and yelled for their dinner at the lady behind the front desk, who, without missing a beat of checking in her campers, waved them toward an empty table in the next room.

Tom and I got a cabin that held all our supplies, didn't require us to climb a 12-foot ladder, and had a covered porch from which, the next afternoon, we saw the clouds gradually clear and heard the drone of an incoming plane. The big, white C-206's pilot's door was already open as it taxied in, strewing plastic bags and empty cans.

As the plane spun to a stop Cade and I exchanged hopeful looks. Extra room on bush planes can be hard to find, but this Skywagon was almost twice the size of the compact tail-dragger we'd flown in on, and even already partly loaded it might be able to take our baggage.

Then we heard a cackle. Canvas pants flapping around his boots, the big Cessna's salty dog pilot strode over.

"So. You're the ones."

"That's right, we're the ones," shot back Cade, one of the few people I'd have no trouble seeing over in a movie theater. "Whatever you think you mean by that."

Not sure if he was being made fun of, Salty looked us up and down.

"Question is," said Cade, "you ever charter?"

"Fuck no. All I haul is freight."

Salty bent to spit a wad of chew, then ground it into the gravel with his boot. "Heard Fish and Wildlife dumped your asses," he chortled. "Wouldn't try Anaktuvuk Pass?"

"Not in a snowstorm," Cade snapped, and with that he turned on his heel and walked away. That meant until somebody else — somebody willing to charter — flew into Bettles, it was up to me.

"Know what?" I said. "We made a mistake."

"All you pussies ever do."

"What we should have done was get ourselves a real bush pilot." I stuck out my hand. "Somebody like . . ."

Salty didn't know what to do but shake.

"Elvin Henderson," he said, breaking into a partially toothed grin. "Pleased to meetcha."

I nodded and started after Cade. "Got to radio for another plane, Elvin," I called. "Maybe we'll run into you. Up on the Slope."

There was a long silence, and I was almost out of earshot when I heard him reply.

"Trip like that," Elvin hollered, "through Anaktuvuk: that'll cost ya."

Two hundred and seventy-three dollars each, cash. And we had to pile our food and equipment on top of the deliveries Henderson already had on board. That filled up every bit of the cargo space behind the two front seats except for a long cavity between the uppermost duffels and the fuselage's metal ceiling, into which Tom volunteered to climb, headfirst. All Elvin and I could see were his feet, but Cade said he was comfortable and could maybe sleep. It was a decision he would regret.

We were heavy on takeoff, but, between splats of chew shot toward a lidless tuna can, Elvin climbed us out. Ahead, the Brooks's peaks were dark, draped with chalky stripes where snowpack filled their drainage furrows, and as Henderson followed the John River upstream, a narrow passage led back into the spotless ice fields of the higher elevations.

As we flew up the little river, I watched the V-shaped walls of Anaktuvuk narrow until there were only a couple of hundred feet between the rock walls. Suddenly, Elvin dipped my wing.

"There's your other plane. Ha! Belly up."

On all sides of the wreck were barren slopes of basalt.

"Know what *Anaktuvuk* means?" he yelled.

Leveling us out between the bluffs, Elvin waited for me to not know, then let out a guffaw.

"*White man shit!* Gold prospectors — Eskimos call whiteys *tuvuk* — came through here. Left so much *anak* — shit — that's what they named the pass."

By now clouds filled the narrow passage, and because we couldn't see a thing Elvin pulled the Skywagon's nose up steeply. Five minutes later the darkness lightened and we ripped out into blinding light. Under a lazuli sky, the crest of the Brooks Range swept past: Sillyasheen Peak to our left, Mount Stuver to our right, both trailing icy plumes blown southward by the polar jet stream pouring over their crests.

Then the mountains were behind, and ahead the olive tundra stretched away to the horizon, speckled with the glaucous splatters of standing pools that flanked a long, twisting channel of silvery water. Elvin nodded in its direction.

" 'At's the Colville. Fish and Wildlife'll be along here."

He spat another wad of Mail Pouch. "You got some place to land?"

Riddle hadn't mentioned how we were supposed to handle that. But next to the river were wide, empty beaches.

"What's wrong with these gravel bars?"

"Ain't gravel." Elvin cupped his hands. "Baby-head rocks." Then he pointed: drawn up on the shore of a bend was Ken and Rebecca's inflatable Zodiac; beside it sat their yellow tent.

"We'll buzz 'em," said Henderson, dropping in. The Riddles waved as we went over, and as Elvin came around after his first pass I saw he wasn't going to climb out after all. We were lined up to land.

"Don't look too bad," Elvin observed as we settled toward the stony shore, and I told myself that whatever happened would be over in a few seconds. Then we slammed down hard, bounced, and as we barreled along, skimming the rocks, simultaneously Elvin and I saw the same thing.

"Fuck!" he screamed, because 50 yards ahead a broad gully crossed the bar. Henderson shoved the throttle forward and grabbed another notch of flaps, but we were too heavy, and with the Skywagon's prop on full throttle, at 40 m.p.h. we went over the gully's edge.

The next thing I knew was the after-impact panic of not being able to breathe, the same thing I'd felt hitting the pavement in four or five motorcycle track-racing crashes. But it never got any less scary, and as I struggled to get air back into my lungs I realized that I was bent forward under the dash, with something heavy and squirming on top of me.

It was Tom Cade, who from his seat belt–less horizontal station had rocketed forward, feet first, into the back of my neck, then continued on into the Sky-wagon's instrument panel. Sloshed with gas from the Cessna's overhead wing tanks and fearing fire, Tom and I frantically pried open the passenger door, yet only after we'd crawled out did I realize that time had passed, since Henderson was yards distant, yelling at us from the safety of a tributary gully to get away from the plane.

Cade and I scrambled over, with Tom using mainly one leg, and flopped on the baby-heads. There was no explosion; besides our gasps there was only silence until, as if in a dream, we heard hoofbeats. I lifted my face from the rocks and peered over the gully's edge at the churning legs and saucer-size hooves of a galloping caribou. Spraying gravel with every step, the heavy chocolate-and-tan bull swerved past Elvin's nose-down Cessna and forded the Colville, throwing up a bow wave of white spray.

I looked at Tom. One hand gripping his lower leg, he shrugged. "Abundant along here. I might have broken my ankle, though."

An hour later Henderson was on top of the wing, securing his Cessna's popped-open fuel-tank hatches, when three Inupiat arrived from the old military base, now a hunting camp, at Umiat. Drawn by the crash, they shook their heads and chuckled, pulling Elvin back and forth around the plane, pointing and laughing louder every time they found a new

317

damage site. None of that ameliorated Henderson's frame of mind, nor did the fact that, for the Inupiat to help him get his big plane up out of the gully the four of them had to unload almost all its cargo.

As the Cessna's long propeller skidded backward out of the corkscrew trench it had dug as we plowed into the cobbled beach, the Inupiat really doubled over, for the terminal foot of each blade was a gnarled stub of aluminum.

Glaring at his helpers Henderson was determined to prevail, and digging out an industrial-size hacksaw he set to work cutting off the twisted ends of his prop. I couldn't believe it. Airplanes were intricate mechanical contrivances — mortally complex craft to be broached only by men in coveralls operating in spotlessly clean hangars lit by overhead fluorescence.

But I was new to Alaska. In five minutes Elvin had shortened his propeller by at least 2 feet, not even taking care to saw off exactly the same amount from each blade. In wonder, I picked up one of the discarded ends and was turning it in my hands when Riddle walked up.

"You reckon he can take that thing off?"

Wide-eyed, I handed him the severed chunk of aluminum. Ken felt its weight and shook his head, but Elvin was already midway through pre-flight. It occurred to me that before his impending crack-up I should get a picture, so I limped over to his open pilot's door, but instead of a smile Elvin yelled, "Fucking asshole," shot out of the plane and smashed my camera onto the rocks. Then he was back aboard, stubby prop spinning, wobbling down the gravel bar on his bent-in right landing strut, whose wheel astonishingly still held a fully inflated tire.

Transfixed, Ken and I watched as Henderson ricocheted from baby-head to baby-head, then pulled his hollow shell of a plane up into a stiff breeze. As he climbed into the distance we both wondered aloud what then, but Cade said Elvin was undoubtedly headed for some private repair strip, where, with new parts, and no photograph to chronicle his mishap, he'd be back in business before the week was out.

A juvenile falcon at Laguna Madre

George (Courtesy of Shawna Fisher)

Alan, still a greenhorn at the flats

Vose Aviation

Three-week-old Crystal, a peregrine/gyrfalcon
hybrid, weighing in (Courtesy of John
Hoolihan)

Peregrine hatchlings in their tub at the Santa
Cruz Predatory Bird Research Group facility

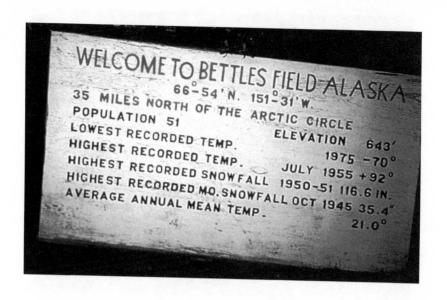

Alaskan welcome

On the wing

323

On a bluff above the Colville, Alan being handed a raptor chick to be banded and photographed

Old whaling boats along
Barrow's Eielson Lagoon

Polar bear on the edge of town

Delgada

Alan and Delgada at Mansfield Channel: she is enraged and hissing as he tries to commit her markings to memory, then releases her (Courtesy of F. Chavez-Ramirez)

A lesser yellowlegs:
typical shorebird prey for the peregrine

Great egret in the Papaloapan marsh

The Martinez brothers with '469 at La Pesca

Under '469's wing strut, the "tall bush" of the
baboon sanctuary along the Belize River

The Belize International Hotel,
known to all as "the Ladyville"

Stolen-nestling parrot salesmen

Mayan temple's upper platform

Out to sea

19.

North Slope

This year I was going back up north for only a few days; maybe, I thought, it would be better not to mention my trip's extension to Jennifer at all. So, with a steadier pilot, I headed for Umiat.

It made Bettles look like Paris. Set 100 yards in from the north bank of the Colville, Umiat — whose sign read POPU-LATION: 1; MAYOR: O. J. SMITH — consisted of a long, pebbly landing strip and a group of old military Quonset huts, a gray-green wood-frame cabin, and several sheet-metal buildings, their roofs festooned with moose antlers, spotlights, and shortwave radio antennas. Upstream from Umiat peregrines nested on the southside river bluffs, as well as downstream on the continuous wall of unclimbable dark clay that for miles bordered the Colville's opposite bank. But right above the airstrip there were falcons in flight, and again this year the river came alive.

Seeing those aerie bluffs, all I could think about was Amelia. No one I'd spoken to in the ornithological community thought it was possible to find her living along even this famous peregrine river, but the same authorities had declared that George and I would never radio-track a peregrine up the Rockies. Nor would they have believed that, after having lost Amelia in the Texas Panhandle, we could have found her again, days later, two thirds of the way to Canada.

It was, of course, also true that even if by some stroke of luck I found her here it would still be of monumentally little consequence. That single data point, measured against the immense annual tides of migrant raptors, would be a thing too statistically insignificant to warrant consideration by most academically oriented students of avian behavior.

Amelia's success or failure was not merely academic to her, however, and it certainly wasn't to me. So in a long aluminum skiff — hired from O.J., a good ol' guy from Llano, Texas, who let me use one of his boats on account of my dad's hunting-lease friendships with some of the same Mason and Llano County ranching and turkey-farming families he also knew

— I set off the couple of dozen miles that were all O.J. was willing to allow his boat to venture upriver, a direction from which I could always row back downstream in case of outboard trouble.

With the receiver's switch on Amelia's frequency, below a bankside ridge I could see a pair of silhouetted lumps that were probably too upright to be rocks, and I nosed in to shore. The scanner was silent, but checking it gave me a second to assess what I'd seen. Huddled in a loose-feathered slump like sleeping owls, these were not the lean, hard-winged peregrines I knew from Padre. And they had not come screaming out at me like nest-defending adults. Already able to fly a bit, they were adolescents who had hopped or flown a little way from their nest scrape, and I could see that they lacked the pallid cheeks and carbon skullcaps that would come in two years. Their dun-colored backs, each mantle feather edged with a pale rim, were familiar from the Texas coast, and I grinned to recognize the streaky, biscuit-colored crowns that defined their race as *tundrius*.

Glowing in the thin arctic sun they seemed wonderful beyond words, but like all young peregrines these two were yelling.

Under the periodic reverberation of their wavering food wails, I set up my spotting scope and scanned the sky. Peregrines have remarkably keen hearing, and I was sure that before long the unremitting "weeee-aaaa, aaa" of the young would draw the old birds back. As their cries — a mixture of piteous begging and the babies' growing anger at the ravenous hunger their parents were evidently no longer willing to completely assuage — grew louder, I grabbed for binoculars.

False alarm. Over the horizon had sailed one of the Alaskan-strain common ravens — huge black predator/scavengers 6 inches longer and nearly a foot broader across the wings than even a big female peregrine. Along with rough-legged hawks, in the Colville Valley ravens nested on the same bluffs as gyrfalcons and peregrines, though they were less common and, according to Cade, did not get along well with peregrines.

For a moment I worried about the ravens' vast appetites. But these young falcons were too large and fiercely taloned for even large corvids to attack, and with the upswell in the cries of its siblings, a third youngster — a tiercel, judging from his slender profile — flapped around from the far side of the bluff.

Never taking her big brown eyes off the sky, the larger female kept up her wailing whine but, fluffed and sullen, did not move. It was normal for her brothers to be more active, since tiercel peregrines mature faster and make their first flight as much as a week before their sisters. (In a reversal of mammalian genetics, where an X chromosome paired with a Y yields a male and two Xs make a female, among birds two Z chromosomes make a male and Z paired with a tiny W chromosome yields a female.) This reversal also often changes the sexual difference in birds' size, and as future egg bearers, female peregrines must be larger. That means they have more growing to do, yet even after both sexes have gained the air females never achieve the same agility as the proportionately longer-winged tiercels. But females are better suited for capturing large prey like waterfowl and ptarmigan, thus effectively partitioning the size range of prey birds taken near their nests with the smaller tiercels, who bring in mostly song- and shorebirds.

This brood was following that pattern, for, with his sister still anchored to her ridge-rock the young male took off over the river. He flew awkwardly but maintained his elevation well past the opposite bank — where my glasses showed what, two minutes before, he had set out toward.

Scissoring upstream were the shimmering wings of another peregrine. Certainly a parent on this defended territory, as its small dark blades drew nearer I could see it was the tiercel, sire of the brood but, at no more than a pound and a half, no bigger than his sons. Compared with them — compared with any other bird — he moved effortlessly, flicking the air with almost absent-minded downward flips of his wing tips.

When he was 50 yards away, the sun caught his plumage and his dusky silhouette burst into color. Below a blue-gray back, his upper breast gleamed creamy white, and under his black cap shone the broad, pale cheeks that distinguish tundra peregrines from the darker-faced races of the lower latitudes. I thought he might try an aerial food transfer, as I'd seen peregrine parents do in California, but even in the air the tiercel was unwilling to risk coming close to his famished son, who swooped awkwardly toward him, talons outstretched for either prey or father. Avoiding his offspring's claws was an easy feint, as the tiercel deftly brought in his slender wings and dropped below his son's roundhouse snatch.

Because the older bird carried no prey, he continued on upriver, leaving his offspring to flap back onto the ridge, still crying. Seeing he was no longer being chased, the tiercel fluttered onto an isolated rock just below the bluff's ridgeline and bobbed his head toward his young, all of whom wailed with renewed vigor at seeing him so close.

He knew what he was doing, though. His having picked a narrow pinnacle prevented any of his brood from landing close enough either to snatch the food he might have brought or to seize his own small body in their claws. Among all three youngsters, satisfying their unrelenting appetites was the sole objective forging through every nerve in their bodies, and armed with hard, scaly toes now able to clench almost as powerfully as adults, the tiercel's progeny had no inhibitions about what they drove their talons into.

After he had settled, fluffed his plumage, and preened, no one moved, and as the sun edged along a violet-gray horizon, a thin film of frost crystallized on the gravel bars and my riverbank tent. But the brightening of dawn was only a couple of hours away, and with the first good light I scanned the scarp just in time to see the female falcon arrive. Grabbing for the radio scanner, I saw that — damn! — she wasn't Amelia. I had promised myself not to hope that every female I saw might be her, but of course I'd forgotten and, feeling silly, got the glasses up just in time to see her alight, using just her right foot, for in her left was something all loose wings and stick-like bill.

Its plumage was more brown than gray, with yellow legs: a snipe, familiar winter resident of the fields near my Central Texas farm. It was strange to find it here in the Arctic, though snipe nested everywhere in the Alaskan lowlands. I got only a glimpse, however, because in a flash of mantling wings its small body was hidden by the cupped remiges of the young female, who, seeing food, had screamed and charged her mother, forcing the older falcon to lift off. Glaring right and left across her umber shoulders, now as broad as those of her parent, the big fledgling hid her food from her siblings, turning her spread tail toward them and, every few bites, driving them back with ferocious screams. I wondered how she could screech and swallow at the same time, but somehow she managed.

That fierceness was why the parents were not willing to come near their offspring, yet even knowing what was going on, it was difficult to reconcile these ferocious adolescents with the helpless newborns cloistered in the hush of the Santa Cruz Predatory Bird Research Group's nursery. There, human volunteers clucked as softly as brooding falcons — clucked even to baby peregrines still asleep in their eggshells. Like doting human parents playing music to infants curled in the womb, at least one of the group's solicitous human nurses had talked with low "chup, chup-chups," right through her charges' shells to the little pink-skinned bodies floating in the sanctuary of their individual amniotic seas.

"E-chup, chup, chup. Pee-chup," she had asked each natal hawk when the time came for it to hatch, her ardent parental clucks imploring it to lift its pipestem neck and, using the big hatch muscle running down its naked nape, jerk its beak upward to pip a pinpoint hole in its confining shell. After that, the baby peregrine would need to rest, sometimes for a couple of days. Eventually, though, with enough encouragement it would manage, using the abrasive material on the end of its snout, to scrape and saw its way around the whole circumference of its shell; then after another long rest it would split the confining halves of its calcium cell and fall out into the world still hidden from its tightly sealed eyes.

Picking up one of those ounce-and-a-quarter nebbishes, Brian Walton, Santa Cruz's director, said that every one of his soggy newborns looked like it needed another month in the shell. But that was only for the first few minutes. Walton's maternity crew knew their job: absorbing the babies' dampness with soft brushes, they fluffed each neonate's initially invisible snowy down, fine as vicuña wool. It was a coat the pampered infants in Walton's towel-draped plastic bins required, because during their first week peregrines' nascent metabolism leaves them nearly as cold-blooded as reptiles, and without a constantly brooding parent they can die from the merest trace of chilly dampness.

Getting the hatchlings even this far had called for the development of complex technology that had been initiated with the incubation of Cadey's first eggs, laid at Cornell in 1972. That clutch was lost in chicken incubators that ran too warm and produced too much humidity for peregrine eggs, so game-bird incubators and maintenance procedures were tried. The process had been fine-tuned over the years by falcon husbandry specialists like Willard Heck, and now the Peregrine Fund and its World Center for Birds of Prey had it right: Walton's pastel-walled, computer-regulated brood room, lined with constantly monitored, clear-topped falcon-specific incubators, suggested nothing so much as a human preemie unit.

Next door, in the nursery, I noticed a puppet stuck on a shelf. It was a famous puppet — a copy of the celebrated peregrine-faced feeding device originally used at Cornell. Its carved wooden face, half again life-size and painstakingly painted with a shiny black cap, an anatomically correct hooked bill, and big glaring eyes, was so ferocious that photographers loved to shoot it poised over the tiniest, Easter-chick-helpless newborn falcon. The idea was that, by reaching through an aperture below the puppet-parent's chin, the fingers of the maternity staff could offer bits of meat, imprinting baby peregrines on a creature that looked at least something like its future self.

Instantly, I wanted to use that puppet, to squeeze cut-up cubes of quail breast into the open mouths of the fuzzy white punkins who all wore wide-eyed expressions of astonishment — and do it harmlessly, never imprinting them on my dangerous human form.

Walton said that was all baloney.

"At first we didn't realize it, but at this stage baby peregrines can't focus on much of anything. You could feed 'em with your shoe."

I used my fingers. Thumb and forefinger propping up the chicks' wobbly necks, using my other hand I tweezered in their tiny feeding bites, chupping solicitously. But only the littlest ones needed help. Peregrines a few days past hatching lifted their heads on their own; a week later, fluffed in longer down, they scrambled across their absorbent-bottomed blue tubs, tipping their pink mouths up to screech, "kee keeeee!" at any shadow that might bring them a bite of quail.

Another week or so, and the eyases' natal fuzz had been replaced with a coat of rough ashen wool, and soon after that some of them had begun to sprout rows of dark, primary-feather quills along the rear of what still looked more like arms than wings. That meant they were almost ready to go.

Not out on their own, but part of the way into the world, via the aeries of wild peregrines that, feeding on the chemical-laced birds found throughout California's agricultural lands, often produced infertile eggs. During darkness, those barren eggs were slipped out of their nest scrapes and replaced with ceramic replicas that the parents invariably accepted — and incubated — until the day of the Big Change.

Then, before dawn, the same climbers who put the substitute eggs in the aerie gently flushed the brooding female off her nest scrape, plucked out the substitute ceramics, and eased a couple of Santa Cruz's big, bin-raised chicks onto the ledge. After that, everybody waits, for it is one of the memorable sights of peregrine study to watch the adults return to find, in place of their cherished, often chupped-to artificial eggs, a pair of hungrily screaming adolescents. Like entering the delivery room and coming out with a bratty thirteen-year-old.

At my designated aerie north of Petaluma, the female falcon had already handled that first morning: she'd been wary of but fascinated with the huge new creatures floundering around her ledge.

For hours she had scrutinized them from a pine snag, then given in to the newcomers' hungry wails and brought them prey.

The tiercel was still distrustful of the two rambunctious beasts, half as big as himself, that had suddenly come into his quiet haven beneath an overhung granite boulder. Most of the morning he sat just above my rudimentary blind, periodically peering in at me — for these were fearless peregrines who every day hunted human-occupied farms and vineyards — and, bobbing his head, tried to get a better view of the scuttling, wing-flapping interlopers who had so enthralled his mate. Yet before long their deafening dependency had also gotten to him, and an hour later he arrived with a freshly killed red-winged blackbird, waiting near the top of the pine for his mate to come and take it to their ledge.

20.

Arktikos

By the next morning, it was clear I'd made a mistake. What I was doing on the Colville was ridiculous. Vose and I had pulled off a couple of miraculously lucky reunions with Amelia. But now I saw that, amid the vast distances of the North Slope to even contemplate finding a single peregrine — even radio-bearing Amelia — was absurd.

Still, she had to be somewhere in the Arctic, and wherever her aerie ledge, by now her brood of young — if in fact she had found her mate and successfully hatched eggs — might also be getting ready to fly. So I decided that during my short time here I'd simply watch as much as I could of these surrogate arctic babies' transition to adulthood.

Yet it seemed I might have been too late, for from the rim of the next high bluff a single, strongly flying adolescent lifted off and disappeared, as had other young peregrines along the river. So with the receiver periodically running its general scan I went on west. From a bar loosely paved with the Colville's ankle-twisting chocolate-brown rocks, well back from the river I spotted an outcropping splattered with mutes.

I'd hoped it might be a nest site, but it was empty and, even covered by droppings, it may have merely served ravens, rough-legs, or non-nesting falcons as an observation roost. Yet from its elevation I scanned the tundra's endless sweep of sky and horizon — a landscape that appeared to operate in a dimension more transparent than other earthly places. With neither water vapor nor dust to soften distant views, the clarity of faraway detail was like that of laboratory-altered, hyper-realist photography, in which details too minute for human eyes are carefully delineated.

The most striking thing about that supranatural clarity, however, was that it occurs in the absence of brilliant sunshine. From the crest of this high bluff the Colville seemed only feet below, running gray-green between grayish gravel bars and cypress-hued dwarf willows. The flinty ridges of the Brooks Range, only seven or eight thousand feet high yet still capped with summer snow, lay on the southern horizon — a strange direction from which to see North America's mountains, almost all of which loom up only from eastern or western aspects.

For all its translucent precision, the land seemed strangely sterile, with only meager vegetation matting its slopes, although beyond the river's banks that olive surface proved astonishingly difficult to traverse. On higher ground, bare reddish soil separated rocks covered with scaly patches of orange, green, gray, and brown lichen. But the low-lying swales were only visually smooth. There the tundra formed a knee-deep mat of stunted birch, cranberry and crowberry, arctic heather, and many kinds of sedge and avens, each plant so thoroughly involved with its neighbors that their meshed stems made up a kind of botanical steel wool too flimsy to walk on and too snarled to bulldoze through.

That was just what grew above the surface. Below its mat of polar plants the tundra was a vast sponge. Everywhere, watery mud sucked in my feet, my ankles, and sometimes my calves — a situation that required waders — although after only a few steps the sedges' broken stems invariably punctured their vulcanized rubber ankles, letting in so much frigid slush that

I might as well have been wearing sandals.

Apart from having perpetually chilled feet, trying to make my way across open ground was like being stuck to a floral tar baby. No sooner had one boot found footing in the under-mud than the need to suck it free called for a robust leg lift. Then that foot had to be waggled free of its tangle of interlaced stems before it could be swung forward, wriggled back down through another wiry mat of sedge, and squished into 6 or 8 more inches of liquid sediment.

After a few dozen strides I felt like Bozo, waddling his oversize dogs around ringside, and after less than an hour I'd lost so much inclination to either pull a boot up or put one down that I simply stood still, a circus Gulliver stitched into immobility by the miniature jungle snarled around his knees.

That, however, wasn't even the worst. The worst was the wide meadows that now blocked my way, filled with tussock-style tundra. The big grassy knobs that filled the Colville's interceding tributary drainages suggested a stunted army of wobble-head dolls: big clumps of loosely anchored heather segregated by watery ravines too wide to stride across and too deep to step down into.

Finally, I gave up and went back to the river. Its interwoven channels were complex but manageable; running rivers was how I'd gotten my start as a tour leader, and on the bars between the Colville's troughs a pair of arctic terns floated up like flakes of airborne ice. Birds of perpetual sun, the newly flighted young that just days before had hidden, as brown-speckled chicks in the variegated gravel that surrounded their nest hollows, were destined to soon leave the tundra. Like their parents they would follow the receding daylight south, all the way to the Andes-backed shores of Peru and Chile, then continue down those rocky beaches to their winter range on the edges of Antarctica.

On one of those secondary channels I motored under the lee of a low rise, sprigged with heather and lichen. Beyond it the air over a marshy slough rippled with the wingbeats of shorebirds — though I saw them only through a haze of mosquitoes. After the last warm spell their swarming larvae had blossomed in every still-water pond, metamorphosing fresh galaxies of ravenous adults that surrounded every warm-blooded creature, clustering in furry rings on the bare-skinned eye rings of songbirds and instantly coating every square inch of epidermis that I accidentally exposed. There, a well-placed slap left a half-dozen little splats of blood, and without my netted hat life would have been impossible; when I took it off the scarlet rash of run-together bites wasn't even the biggest problem.

Breathing was, since my nostrils, swollen from the bites inside, would then clog further with the legions of mosquitoes still seeking the blood of close-to-the-surface nasal capillaries. In the mornings, pursuing the same warmth, other battalions coated my breakfast cereal. Seconds after it came off the primus stove every millimeter of oatmeal would be skimmed with a wriggling fleece of mired mosquitoes. Scraping them off only allowed new hordes to fling themselves onto the path I'd just shaved with my spoon, and eventually I simply shoveled down their hairy ranks along with the underlying oats.

Near the end of this chalky abutment was the big stick nest of rough-legged hawks, which often nest in proximity to falcons. Then I heard the angry chatter of a female peregrine. She was close enough for me to see her shoulders bulge as she stooped at me, then arced her wings to pull up, screeching. She wouldn't have behaved that way if there weren't an aerie nearby, so to lower my threat profile I squatted by my telescope's tripod and watched as she came back around, showing me the light-gray stippling of her underwings and lower breast and the pale yellow V her legs scribed against the wavy crossbands of her tail.

"Amelia?"

The scanner hummed in silence, of course, and when I looked up from its little digital window the falcon had lifted clear of the river and, her angry chatter stifled by distance, was circling upward. I followed her progress until, far below, there was a flick of wings over the bluff. Another peregrine — her tiercel, I guessed — slid behind a mound of grass at the edge of a depression where erosion had notched the slope of a big dirt bluff.

Almost instantly the tiercel was back in the air. Coupled with his mate's anger at finding me so near, his swift departure might mean nestlings: eyases too big for their dad to risk being grabbed by them but still not old enough to be perched out on the rocks.

It could be a lucky find. I took a long time working my way up the side of the ridge, breathing through a bandanna mosquito filter. After a rest I started down the face of the slope, and there, behind the grass where the tiercel had landed, sat three well-feathered, biscuit-crowned young peregrines. Except for patches of baby-wool they looked almost like adults, cheeks plastered with the dark mustachial bars that gave them the ferocious glare of grown-up falcons.

Four-plus weeks: right on the brink of flight, though they had not yet known the air or they wouldn't still have been huddled in front of me. Maybe this was a group I could watch as they found their wings — and do so without causing them danger, for in order to escape an intruder, flightless younger peregrines sometimes fling themselves right off their cliffside nest scrapes. But these kids were well beyond that stage, glaring with gigantic eyes past beaks flared wide in silent threat. It was mostly sham, but to avoid disturbing them further I eased back from their shallow dugout to watch from a distance.

The sun was lower by the time I made it down the bluff, in whose shelter grew willows. Among their forks, redpolls — brown-streaked finches with dark-pink foreheads — hung like chickadees, probing buds and branching twigs for insects. They leapt away in undulating flight at my approach, whistling their hard, trilling cries. But I was far away in thought.

Any day now those young peregrines would steel themselves for their first plunge into the void. I wondered if it would be a struggle — but that was as far as I got, because in an opening among the trees lay a mound of droppings that could have come from only one creature.

Shaking, I repeated to myself what I'd heard in some bar in Fairbanks: "Steamin' bear shit: That's when you grab a tree."

I ran my fingers over the inch-thick trunks of the largest surrounding willows — saplings that stretched up 10 or 12 spindly feet. But the grizzly's tracks were so old and hard they were just scratch marks, and I took a breath. Then I remembered reading about some Alaskan biologist who'd found in a pile of brown bear excrement a whole sardine can. Unopened.

Cautiously I poked through the gigantic, surprisingly human-like dung. Good and dry; no unopened cans. But I was scared: a hostile encounter with a bear was unlikely, but still possible. I had pepper spray and my dad's old .45 automatic, but here on the northern edge of the continent a barren-ground grizzly might go through all its years of adolescence and maturity and reach old age without ever seeing a human being. Having no innate fear of any other animal, bears like that, I'd heard, sometimes just wandered into camp. And with grizzlies there was no telling what might happen then, because a bear's sense of smell is so keen that it is impossible to live immaculately enough for one's belongings to be free of the aroma of good things to eat.

On the Arctic Slope you didn't even have to live in wilderness to run into one of Alaska's 30,000 grizzlies: half an hour's bush-plane flight north of here a 700-pound bear had walked through an open side door of the Prudhoe Bay Hotel, climbed a set of stairs to the second floor, and, before she was shot, had begun nosing her way into rooms occupied by offshore drilling workers.

My campsite — with canned food, provisions that included not just sardines but tuna and (bears' favorite) salmon — was right on the other side of the river. I didn't want to keep standing there by the scat, but I didn't want to go back to my tent, either. What if the grizzly showed up while I was asleep? What if I could hear it snuffling around, looking for my cans of salmon?

I knew exactly what I'd do. Once lions had come in the night to a Serengeti water hole where I had set up camp after my tour clients had gone home. I could hear the rumble of the cats' great bellies and the garden-hose gush and acrid stench of the males' territorial urine, so I did the only — entirely useless — thing I could. I lay perfectly still, exactly centered in my small tent, maximizing the 2 feet of space separating me on each side from the thin nylon walls that were all that stood between me and those terrible, nearby claws.

That evening I didn't heat my food — canned beans and corn only — and for fear of dispersing even their mild vegetable scent I washed and buried the cans, deeply, on the far side of the river. Heavy rain — which I feared even more than grizzlies because I knew how quickly hypothermia could kill you here, even in summer — had held off so far. But a chilling haze had drifted in from the sea, hushing birdsong and roofing the valley from the sky. Under the vapor's low canopy, coveys of sandpipers wove the darting little flights that dropped them precipitously into bunched roosts along the water's edge, and from the marshy ground beyond the shore came the secretive titter of golden plover.

Yet, unseen beyond the clouds, a peregrine orbited over this calm world. Still calling, two of the plover rose above the misting river, and even in the failing light I could see the honey and charcoal of their backs pulse with every wingbeat. Beyond the intervening haze the peregrine also saw that glint and came down, knifing silently through the fog's sheltering vapor. In an instant, he was on us.

From their height the plover saw the gravel below them fling off its pale rime of shorebirds, saw twenty sandpiper wings gust into flight, and knew they were in danger. The falcon's gunsight eyes had locked on the pair, but when he arrowed in both plover twisted onto their sides to shrink the black feathers of their breasts away from his outstretched talons.

As he flashed past, the tiercel was lost in a pandemonium of rising shorebirds, but hidden by their streaming wings he threw himself up from his dive and rose again toward his prey. For the plover, safety lay in the sheltering sedge, and as the darker-plumaged male dropped toward its thicket of stems, the falcon met him coming up, rolled onto his back, and with his fist of claws reached out and seized the plover's breast with crushing force. Ten feet beyond her mate's crumpled body the female plover skittered away. Accounts from the old market-hunting days tell how shorebirds sometimes circled back, crying, over a gun-shot companion. But a bird caught by a peregrine is always alone. No bond is strong enough to overcome the visceral terror that drives its comrades to blindly flee the clutch of a falcon's claws, and in her ancient panic the female spared not a backward glance.

I felt for her, and for her mate, because in that brief swirl of dark wings the life suddenly stilled was one whose heroic contours matched the peregrine's own. Within that slain plover lay both a global map and, perhaps unmatched on earth, the will to use it. In less than three weeks, this diminutive seeker of worms and snails, picker of gooseberries, and foot-tapping listener for the subterranean whisper of hidden amphipods and larval insects would have made his way across 2,000 miles of arctic tundra and eastern boreal forest to Canada's Maritime Provinces.

From there, almost without pause he would have launched his 4-ounce body out over the North Atlantic — an ocean until recently uncrossable by air for humans other than long-distance heroes like Lindbergh and Beryl Markham. Over that realm of trackless waves, along with others of his kind this plover might have climbed to 20,000 feet. There, wings beating four times a second, he would have forged on 3,000 miles to Surinam, on the northern coast of South America, making one of the longest open-water flights any creature has ever accomplished. And having once again reached land, 20 percent lighter from the effort of his crossing, he still would have gone on, winging his way over 1,000 more miles of tropical rain forest to the Atlantic coast of Brazil.

21.

Cherokee

It was a death hard to ignore. My own life, it was clear, was worth no more and probably far less than that of any of these small, valiant beings: creatures I saw, every day, escaping danger or having their existences casually consumed by some other animal. Lives that were usually lost not because of some elemental mistake in an individual's tactics of survival, some momentary oversight, forgetfulness, or imperfect adaptation to the environment. Lives lost, mostly, just to chance.

Evolutionary theory notwithstanding, a majority of these wild creatures seemed to live or die by simple luck. Random life or death was what faced almost every warbler, shorebird, wagtail, and sparrow. Like the golden plover, most of the prey birds taken by the Colville peregrines seemed to have simply found themselves vulnerable at the wrong time.

Most members of fast-reproducing,

heavily preyed upon species die in a predator's jaws or talons even if they are as careful, highly vigilant, and ungreedy as their genes allow. Even if they do absolutely nothing wrong. And my existence out here was no different from theirs. Even close to Umiat, on the shore of this remote northern river a grizzly, drawn by my scent or that of my food — maybe one who simply happened to be wandering past — could amble up. If it was in a bad frame of mind and happened to attack, it wouldn't be establishing some proper balance by pruning the weak or unwary; 200 miles north of the nearest whiff of political/ecological correctness it would kill me simply because I was there.

I twisted my neck to scan the gravel bar, where at least I could spot an approaching bear a few yards off. Not that it would do me any good. But in the shadows, every aberrant gurgle of the river had become a grizzly's snort, and with a pounding chest I'd imagine yellow claws the size of my fingers raking the rocks where my washed-out cans were buried. Then a long, coarse snout would lift into the wind and turn toward my camp across the shallow channel. . . .

In the morning, outside the tent I found

a row of tracks. Huge, but, thank God, not grizzly. They were made by hooves, nearly as wide as those of a dairy bull but split, like the prints of hogs, goats, and sheep. Whatever had left them was heavy: each track was sunk so deeply into the mud between the shoreline rocks that its depression had acquired a little oval pond of seepage.

Moose. A cow, as the females are called, although moose are really giant deer and ought to be referred to as bucks and does. Surely yesterday I'd have noticed her trail. Yet, heavy as a horse, without my realizing it she had stalked across the pale-seamed rocks 20 feet from where I slept.

So much for listening for a bear.

But in the calm morning light both moose and grizzly seemed almost imaginary, and as I glassed the lip of the invisible aerie I slowly realized that the whitish flecks floating above the nest site were feathers. That meant the young already had a kill. So, detouring around yesterday's saplings where the grizzly dung lay hidden, I climbed through clouds of mosquitoes still sluggish from the nighttime chill.

From the top of the ridge it was clear that the pale feathers were also barred, marked with both tan and dark brown.

Drifted along the face of the bluff, they were breast and flank plumage, the feathers peregrines typically pluck away first. Though they were soft and fluffy in my fingers, they lacked the filamentous down that surrounds the lower quills of waterfowl: what gave them their loft was a frizzy sub-feather that branched off the main shaft just above the bird's skin.

That meant they were feathers from a grouse; in northern Alaska, ptarmigan, and a glimpse through the grass fringing the aerie showed there wasn't much but feathers left to identify from beneath the mantling wings and tail of the smallest and probably last-to-feed adolescent. In the silence I could hear the rasp of the tiercel's beak on hollow bone and the frustrated hisses with which he scraped at the shell of lower legs, feet, and plumage.

By evening, that morning's three quarters of a pound of ptarmigan, split two and a half ways, was forgotten, and every one of the brood was once more keen with hunger. For the first time they had moved out onto the rocks to send forth the wavering wails their empty crops forced from hungry throats. They couldn't have been more obvious to a predator, and I worried a bit for their safety.

But they were as large as grown falcons now, with hooked black talons too formidable for either fox or jaeger. Snowy owls mostly stayed downstream, toward the coast, and while golden eagles sometimes took peregrine chicks, I hadn't seen any. Then, just before dark the parent tiercel came in with prey. That sent the young into a cacophony of wails, but their father wasn't moved. Back and forth along the bluff, he fluttered his wings like a long-limbed moth, dangling a shorebird from one down-stretched foot. As he flew, he imitated the food cry of his young, and all three waved their equally lengthy wings and piteously wailed back. But none of them released their hold on the rocks, and after his third pass the tiercel flew to the top of the ridge and plumed and ate his sandpiper in five minutes, then fluffed his feathers for the night.

It was time for me to turn in, too, but I couldn't sleep. It wasn't the intrusion of the almost-midnight sun, for by now a few hours of darkness had returned to the Arctic, though the shadows only made it worse because in their dimness, I felt sure, lurked bears. To keep from thinking about them I pictured the adolescent peregrines on their bluff across the river and tried to imagine the transformation they were about to undergo.

That brought back Cherokee.

Once they have flown even a little, raptors' souls wed themselves to the air in ways probably impossible for humans to comprehend — bonds forged with such intensity that their expression can be agonizing to watch. Gravely wounded, Cherokee had been brought to the Austin Natural Science Center, where I worked. A just-fledged red-tailed hawk still patched with milky down, he had been an easy target for some scumbag with a shotgun who'd blown apart his left wing, right up to the shoulder.

When I first saw him, his entire humerus was a twisted, open wound seething with maggots. Maggots in the lacerations of injured wildlife always get to people, though they may not actually be so bad because fly larvae eat only dead flesh, and (though this is debatable) by ridding the area of necrotic tissue they may actually eliminate some of the sepsis that would otherwise occur.

But maggots are still a sign of severe, untended muscular damage, and Cherokee had more of them than I'd ever seen on an animal his size. After I'd cleaned his wound and hydrated and fed him — he was such a baby he squealed like a nestling when he saw a just-thawed mouse on its way to his beak — there was nothing to do but, using my rudimentary veterinary training, amputate the twisted wing he'd already slung around so violently its shattered bones had torn into his shoulder.

Eventually, however, Cherokee healed and got his Indian name from the kids who used the feathers from his severed wing for headdresses. Then, because the Center always had more hurt hawks than it could support, Cherokee came home with me, and from him I learned the power of the sky.

Inside my backyard's tall cedar fence he'd tear around on foot, holding out his good wing for balance, and leap up onto my wrist for his daily mice. But he never forgot what lay above. From my arm he'd cock a burning eye upward at — as far as I could tell — nothing.

Then there'd be the faint contrail of a passing jet, and I'd figure Cherokee was scrutinizing the passengers in its windows. But he didn't just watch the sky. He wanted to be part of it. Midway through his first year he learned to coordinate his jumps with frantic flaps from his strong right wing. That let him reach the first limb of a big oak, from which he'd hop and flap all the way to its topmost branch.

There he would spend the day: heart in the clouds, riveting every passing jay or overflying vulture with a gleaming golden eye. But it was not enough. Every few hours the freedom — maybe just the movement — he saw in other flying things would grow too much, and Cherokee would clench his straw-colored feet, squat, and, from his worn-bare branch, fling himself upward, flapping with all his strength. Face turned hopefully toward the firmament, he'd gain a yard or two on earth before it pulled him back, parachuting down onto the grass. Before dark, though, he'd be back on his perch, ready to try again, which I let him do because he was clearly less alive when I kept him in.

But here's the thing: this naïve baby, who might have spent no more than a day on the wing before he was shot, then spent *twelve years* trying to regain the sky.

What was difficult to understand was why the sleek, fully feathered adolescent peregrines on the opposite bluff — even as ravenous as wolves and with a prey bird dragged in front of them — only wailed and flapped. Maybe Cherokee's obsession with flight would come later, after some mysterious switch had closed, completing a primal arc that, once and for all, would make these fledglings creatures of another dimension. That was the kind of thing George said had happened to him the day the barnstormer came to town, and I fell asleep with my falcons fading into a yellow Eagle Rock stunter going round and round and vertically round like a golden Ferris wheel, spinning its magic over a small-town schoolhouse in Machias, Maine, sixty years in the past.

By morning, all three siblings had moved away from the aerie. As I moved the telescope from one to the other, I saw that, despite being even hungrier, today they seemed less helpless, and through the lens I could see each youngster's umber nape ruff out with excitement at the sight of every far-off rough-legged hawk or glaucous gull. Then all the young falcon's body feathers flared, her screams went up a notch, and when the parent tiercel suddenly swirled across the top of the bluff carrying another lump of passerine beneath his tail, the two juvenile males leapt stiff-pinioned into the air to meet him.

It was almost a disappointment. After these babies' near-desperate determination not to leave their cliffside perches, it was astonishing to see them almost casually set off on the wing. Their nonchalant launches seemed to take no internal steeling, no rearranging of their basic circuitry: when they were hungry enough and food was in sight, the fledglings simply spread their arms and stepped out onto the air.

But for a while they *had* been afraid. To survive, for weeks young raptors must dread the encircling abyss. Yet even with that fear, in each nestling peregrine there was also something of Cherokee's mad stare. As soon as its eyes could see, every white-downed eyas had gazed for hours at the overarching blue; that was where its parents and its food came from, but even when its crop was full, its eyes had still watched the sky.

Moving downriver, within 5 or 6 miles I saw another falcon, and as I watched this circling female's narrower-winged mate gyre up to join her, I thought that meant there would be another brood; arriving beneath the nearest cliff, I expected both adults to come down at me. Instead, in the warm air rising from the headland both hawks climbed away from the water, turning to stay within a small thermal, and soon they had shrunk to flecks that, together, disappeared into the blue.

I wondered if they could be passage peregrines, nomads stopping here only for a time, but several perches were so painted with mutes that they must have been used all summer. Located below a rocky promontory, the most heavily covered spot would have been inaccessible to a terrestrial predator, yet for hours its inhabitants did not return, and late in the afternoon, from her broad wings I again picked out the female, circling in the same flat arcs that Amelia had used during her last days in Texas.

Maybe this falcon was old. Non-breeding peregrines sometimes returned to their ancestral aeries, although those without offspring often moved on in the face of competition from younger falcons. Or these two could have tried to nest and failed, defeated by the internal stew of organochlorines their innocent wings might have carried thousands of miles here, to burn away the life of their eggs.

During the 1960s and 1970s, after peregrines had vanished from the eastern seaboard, solitary pairs, afflicted by the chemical blend of which DDT was the first identified component, were reported throughout Europe and the southern Rockies, sometimes defending traditional nest sites empty of eggs or young. In 1971 Cade reported that even the Colville peregrines were "severely burdened with high levels of DDT residues. They produced thin-shelled eggs," he added, "like those that characterized the devastated populations of Britain, the eastern United States, and California because eggshell fragments from 12 of the Colville nests had [been found to] average shell thicknesses . . . 31% below the pre-DDT average."

Hoping for shells that might tell whether this pair had lain defective eggs, I considered trying, the next morning, to reach their aerie even without a climbing partner. Instead, I awoke to a gale. It was the first of the late-summer storms that within weeks would push inland across the Arctic Slope's low plain, sweeping streams of migrants southward through the Brooks Range passes. On the river it was miserable. Eyes squeezed to slits, sparrows and thrushes, wagtails and longspurs crouched under the sedge, swept by the high-pressure gusts that scoured its stems. Even the redpolls — bad-weather birds — would not fly. At the downstream end of my rocky beach, five miniature plover — semi-palmateds, a tight flock made up mostly of that season's young — squatted almost flat when a wind blast scattered the pebbles around their feet, but the mosquitoes were gone. Frozen into chitinous specks and blown south, I hoped, past the mountains, past Fairbanks, and to Hell and gone and good riddance. With my blessing.

Aside from that happy circumstance, every other living thing I could see hated the tempest. Everything but the falcons.

In the dark gusts above the river, both of the bluff's resident peregrines were colorless: sharp-edged cutouts jerking right and left like Gamelan puppets. Transformed, the languorous soarers of the previous afternoon chopped the air with fervent strokes, surging up the wind that howled along their valley. Through the glasses I saw the tiercel climb on excited wings, then let himself fall backward, head cocked onto his nape, looking at the ground completely upside down.

At 300 feet, he saw his mate below, partly opened one wing to roll himself over, and forging his slender body into a feathered spear point, launched a stoop directly at her. It seemed he might impale the female's floating back, but as he came down the falcon peregrine turned belly up, thrust out her talons, and clutched at the ball of clenched toes he flashed by as he leveled out. Instantly, the falcon then folded her wings, fell 50 feet, reopened them to sweep upward, rising beneath the tiercel as she would below a prey bird unable to escape her relentless climb.

He looked back, saw her coming up and screamed, and for the first time I recognized the barbarous joy that flowed through every beat of their wings. It was nothing like the jovial play of the ravens rollicking in the orographic flows of Capulin volcano. Jolted by unseen blasts, yet savage as the gale itself, this pair swept along like distressed kites, buffeted up onto one wing or the other and finally — the incarnation of everything wild and unattainable — gusting exultantly away past the bluff's far end.

Drained, I sank onto the rocks. I could not have seen more. But like everything else about peregrines, their flight in the windstorm — impossible for any other flying creature and too intense even for me to watch for long — must also serve their lives in other ways. Mostly solitary wanderers, tundra falcons are believed to spend only a few weeks together every year. But somehow that is enough. Enough to crystallize, even in childless pairs like this, some long-lasting concept of mutuality, some permanent attachment beyond the shared memory of incubation and chick-rearing.

Of course there were other possibilities. These two might not have been together long. One or the other could be a mid-season replacement, filling the niche left by the other's fallen mate. But even if that were so, this bluff was likely to remain their mutual home. Though peregrines are not thought to either migrate or winter together, as a bonded pair, after making their long, separate journeys south these two might return to occupy this aerie for five, six, even ten consecutive seasons.

That meant their recollection of these last few summer days — days spent soaring high, silent pirouettes — would somehow unite them during the long months and thousands of miles they were about to spend apart. Only that, and perhaps the way they'd shared the storm, would have the power to pull them back next year, during the bright, cold days of May, from half a world away, home.

★ ★ ★

In the morning, it was sunny and without wind, but a deeper chill told how soon the remaining warmth would go. That would bring hunters, because, as wild as it seemed, the Colville was not an untouched river. On its bars I'd seen the remains of campfires, and three Inupiat youths in a motor canoe had come by on their way downriver.

Away from the water were game trails, their surfaces embossed with the circular prints of caribou splayed like doubled commas, front hooves' dewclaws pegged alongside their hind-hoof tracks. Farther in grew saxifrage. Adapted to relentless wind, like other arctic plants it stays low, trailing its stems and glossy leaves along the ground. Native fishermen and their families living in summer bivouacs along the river sometimes ate its blooms when berries were scarce, and the vegetation — dug from the snow by the black-tipped claws I'd noticed on the peregrines' ptarmigan carcass — sustains the arctic grouse during sub-zero winters.

But avoiding the wind wasn't the only reason the tundra's vegetation was so small. Minute sedges, reindeer lichens, moss campions, pygmy buttercups, mastodon flowers, and mountain avens — some older than I — were compactly fitted into crevices between the frozen rocks, sending out barely a tendril a season, yet so vulnerable that with every step beyond the banks I crushed a dozen minuscule leaves and stems. This was vegetation designed to contend with the elements. Caribou and musk oxen concentrated on the valleys' lichens; on the higher slopes, plants' priority was to make the best of the seasonal brevity of sunlight, the lack of nutrients in the thin, acidic soil, and the difficulty in accessing the Arctic's minimal precipitation, most of which spends the majority of the year locked up as ice.

Between their stems I was also looking for ice. Riddle had told me about permafrost lenses, spots along the river where ice seams bowed to the surface — the glassy summits of subterranean waves that curved down to join the frigid reservoir below — for in these lands beside the polar sea, under everything there is always ice. Water frozen for millennia, it is ice too deep to feel the summer thaw — permanent frost that in places goes down hundreds of feet and has not seen the sun since its moisture, at least some of it, dripped as Pleistocene rain or snow from the backs of ancient grizzlies, woolly ancestral horses, camels, and shaggy arctic elephants, all of whose bodies lie entombed within the frozen crypt below.

I'd seen a little of that vanished world. The year before, with the Riddles and Tom Cade I had come to a cliff whose slabs of matrixed gravel periodically thundered into the river. As Cade and I approached the high bank, called Booming Bluff for the concussive impacts audible a half mile upstream as summer snowmelt undercut its base, we noticed something was out of place. Far up the steeply inclined slope — from yellowish clay that had never grown a tree — projected a pale stump.

Up close, that tubular, sand-colored stub became a tusk. Cade said that from their helicopters, oil-company pilots spotted mammoth ivory along here all the time, but, smitten nevertheless, Ken and I scrambled up the cliff. Just below the crest we came to a stop.

Jutting from the crumbly chalk was the broken tusk and part of the jaw of a woolly mammoth. It was not a fossil — a cold, lithic mold of what had been — but once-flesh-covered real bone: a frozen body left from a time when in Europe mankind wore the heavy brows and massive limbs of Neanderthal.

Astonished, Ken and I gazed down. We felt like Lewis and Clark, and I remembered that Jefferson had in fact ordered Clark to bring back a living mammoth, which he fully expected the expedition to find still grazing along the upper Missouri. The great beast at our feet would have stood more than 10 feet at the shoulder, 12 or 13 if it was a bull, and could have weighed 8 tons. But Cade was yelling for Ken and me to get off the bluff before it calved, so with Riddle I tugged out the short section of tusk, whose curled tip had already broken off. While Tom and I got that hunk of ivory on board, Ken climbed back up, then skidded down like a running back with one of the mammoth's football-size molars tucked under his arm.

In camp, it was as if we had exhumed a mummy; during the few hours the tusk had lain exposed, its ancient cementing protein had begun to oxidize. Beneath a dark mineral overlay of vivianite, typically found on the bones of Pleistocene mammals, long cracks had worked their way deep into the tusk's ever-more-brittle bone. Yet as I ran my fingers along its eroding surface I pictured the smooth white ivory the living elephant had carried.

He had been a grazing beast, an inhabitant not of tundra but of a rich tallgrass prairie that even during the peak of the great glaciers had remained free of permanent ice. But then, as now, snow covered the ground for much of the year, and to reach the forage beneath it mammoths evolved the world's largest snow-removal system. Beyond the thick swell of our tusk an elegant compound curve had tapered out to a hooked tip that would have paralleled the ground when the shaggy elephant lowered his head. There, powered by the mammoth's big, columnar legs, those doubled ivory booms would have swept a 6-foot swath clear of snow, right down to the edible grass.

Tusks were not used for defense — mammoths' safety lay in their size — but their long ivory teeth were marvelous natural architecture, functional even after the mammoths' death, for on the steppes of the Ukraine they had been put to an equally ingenious human use. Near Kiev, Soviet archaeologists have excavated houses built by the first modern people to move into the region's frigid grasslands. They were not bands of hunters, at least not of game as formidable as mammoths, but for the framework of their shelters they used the woolly elephants' skulls and the tusks they found scattered on the plain.

Turned upside down, those easy-chair-size craniums made a circular foundation 12 feet across, inside which the ancient elephants' bulbous foreheads provided comfortable seats that tilted back just enough for me, sitting inside one such rebuilt structure, to look up at its smoothly bowed rafters, which consisted of the mammoths' 10-foot-long tusks. Set into the skulls' upturned sockets and rotated until every bent-in tip joined at the roof's low apex, they once held up the reindeer and bison skins that had formed the walls and ceilings of our ancestors' first substantial open-country dwellings.

22.

Alone

This year I would not see Booming Bluff; for me, as for every summer creature, time was running out. Coming back, not far upstream from Umiat's nearby falcon bluffs I was already glassing the air over the river, but as I drew closer no parental chatter scolded, no angry wings swept out, and only when I trained the telescope on the ridge did I see, against the sky, a single peregrine. It was another juvenile, slouched in hungry sullenness, for when the wind shifted I could hear its wails for food.

Sometimes, when a river-crossing passerine slipped by the youngster would plummet, though always far too late. During the next two hours she dove at a sandpiper and twice at redpolls flitting above the willows. Never in danger, the sandpiper easily outflew her, while the finches ducked back into the cover of their branches long before the adolescent, crying with frustration, swept past. It was hard to imagine that her parents were not just hanging back, watching to see if their offspring could survive without them, because until now every ravenous adolescent had been eating — small birds or parts of larger ones — several times a day, and even with an occasional lucky kill none of them seemed able to capture nearly that much prey on their own.

Now that I, too, had to leave the river, I tried to think what might have gone wrong. Their parents had dragged slain birds through the air to lure these juveniles off their rocky perches, then dropped fresh carcasses to hone their flight. But I hadn't expected the adults to abandon offspring who were still without the speed or skill to capture their own prey.

And even with their stored reserves of fat there was not much time to learn, for already there were fewer small birds along the river. Most were also youngsters whose parents had already migrated, but they were quick and agile and supremely difficult to hunt. Like the juvenile falcons, every one of them would soon face the floating ice that would come sliding downriver, and the frost and snow that would cover the tundra's forage. By the next afternoon I realized I would never know what happened: my last adolescent had disappeared.

Maybe its parents were nearby, out on the tundra, waiting. Young arctic peregrines had been seen there, still being fed by their elders, but for how long it was not known. And even if the young had followed the older birds through the mountain passes, long before they reached the northern end of the Great Plains it was thought that every juvenile would be entirely on its own, facing the seemingly impossible task of capturing the 5 or 6 ounces of avian flesh it required every day.

But staying alive was not the only challenge these almost-babies faced. Simultaneously — because with tundra falcons there seemed to be no passing on of the wisdom of generations — each newly flighted youngster was also embarked on a journey of almost unimaginable proportions, seeking a distant wintering ground that lay as much as half the planet away. How other arctic offspring made that immense voyage I had no idea. But I felt sure there was no way the clumsy post-fledglings I'd been watching — baby falcons who in their entire brief lives had probably flown less than 50 miles — could do it, and for a moment the thought swept over me that if my youngsters hadn't all vanished I'd have gathered them up and taken them south on the plane with me.

Back at Umiat, the population had exploded: the three young Inupiat I'd seen on the river had come in, and as I waited by the airstrip they sidled up to ask if I had any painkillers or cocaine. Instead, I gave them what was left of my provisions, including ten scrupulously unopened cans of salmon and sardines that I told them to be careful with — even sealed — up around what was now, to me at least, Grizzly Bluff.

An hour later, at Deadhorse, the land around the airport was nothing like the way I'd seen it before. The previous July, beside the Beaufort Sea that bright green cottongrass plain had been a mid-summer idyll, dappled under a cloudless sky with caribou does and fawns. Now, most of the herd had vanished, driven inland by warble and nose-bot flies, and over the brownish turf flowed ashen clouds that intimated both winter and the loss of my still-unready young peregrines.

Not like the geese that, all along this coast, were also getting ready to go. Pushed by a wind straight off the polar ice, every day the flocks came in, some of them lesser snows from island breeding grounds up in Canadian Nunavut; most of the white-fronts gathered from nearer nesting territories all over the Chukchi Plain. Filling the air with the roar of their throats and wings, above me 2,000 white-fronts circled among a scattering of snows. From horizon to horizon every cubic yard of air seemed to be occupied by a flying goose, and in the excitement of their numbers they joined voices in waves of sound that pulsed louder as new groups arrived or others leapt into flight.

What the geese were sharing was a common belief in their own strength — in the prospect of riding their wings south to a different world, then having the power to return. That visceral conviction was the cornerstone of their ethology, but what seemed to be their single, uniform roar was nothing of the kind.

To its makers, every bit of that vast din was the conversation of families. Grouped in enormous throngs, geese nevertheless travel, for the most part, in domestic units. Though it seemed like every individual on the Deadhorse plain was constantly in full cry, long-range lenses and fast film have shown that in each flock fewer than one of every dozen geese is calling at any time.

And it is almost always doing so for a reason, because even in the midst of their huge aggregations each goose family — parents and their young of the season, who may stay together for as much as a year — keeps in touch by calling out to one another in voices they recognize from many yards away.

It's a good system. Nearly every house in our Houston neighborhood had a different whistle to bring in its kids, and in the summer dusk half a dozen distinctive warbles would float over driveways and across cyclone fences, drawing us to our separate dinner tables. The same sort of brood call lets goose families travel together for thousands of miles, the young encouraged every flap of the way by one another's company and their parents' verbal support.

For the older birds this is much more than merely getting their offspring south. Every autumn, each filial nucleus is transmitting culture: passing on the ancient knowledge of how to live through the winter. For the young, most of that knowhow entails memorizing topography — learning the traditional aerial highways still in use after millennia. But it also means much more. Juveniles don't simply dog the flight of their elders; they absorb their parents' active teaching — instruction that also gives them the alternate flight paths, stopover feeding sites, sleeping roosts, and seasonal schedules they have to master by the time the older birds are no longer there to guide them. When tumultuous weather breaks up their families, though, few things are sadder than the displaced young, birds like the juvenile lesser snow that for days waded back and forth through a rainwater parking-lot pond near my house, honking futilely at the empty sky.

I knew that the old birds above me now could tell as well as I, with my weather radio, what was building behind the clouds rolling in from the sea. It was time, they called, and as their offspring leapt up to join them, their shimmering banners rose higher, merging with the icy haze, leaving behind only the ripple of their departed wings.

As a boy on the Gulf Coast prairies where many of these Deadhorse geese were going, I had waited for them and for the cranes, though the geese always arrived first, embedded in swells of the arctic air that had carried them south. There, those families of snows and white-fronts had released the grip their cupped wings held on the wind, and as I watched, came dropping out of the wet October clouds, tumbling like petals scattered onto the furrows of corn stubble and sorghum.

Up close, there was nothing blossom-like about the geese. They were big and loud with hunger — and, for the old ones, the excitement of recognition and return. Jostling and scurrying over the fallen stalks, each family seemed to regroup, squabbling for forage as the young lowered their snaky necks to scoop up their first taste of the coastal prairie's fallen grain. For the next four months, the geese were home.

It was next to the same plowed fields that, thirty years later, I'd stopped my truck to watch a bald eagle go over the resting flocks. He had swung down from some invisible height — not stooped, like a peregrine, just slanted in on a long, falling angle that brought him across the dark fields, churning into the air a pale wake of wing-thrashing geese.

The eagle had come to the 20 square miles of long-grain rice that made up the Duncan Plantation, as he did every day after the gunners went home, because it was the easiest way to hunt. At dawn, big-bore shells had shaken the rice paddies, leaving wounded geese hidden in the reeds along the bordering dikes, but later the cripples always crept back among the hungry bands that returned to the fields as soon as the firing stopped. The eagle knew the injured geese were there, and as soon as he had swept an earthen swath clear of those who could fly he turned back, seeking the pale shapes he had left exposed on the empty land.

In a cold, biological way it was the best solution for the legacy of pain the hunters had left. But I wasn't a biologist, and in the center of the paddy nearest me a juvenile white-front had fallen, not only robbed of flight but barely able to keep his head above the 6 inches of opaque water where most of his body lay submerged.

The underlying mud was deeper than I expected, and as his head dipped toward the surface I broke into a splashing jog. I got there just in time to lift his beak and haul him up: a heavy 10 pounds, counting his body's 4 or 5 plus at least as much water weighing down his bedraggled plumage. As we headed back toward the road, on silent wings three immature white-fronts and a single adult pulsed by, looking down.

Chances were, they were not his family. But, safe on higher ground, I could see that too many pellets had gone in too deeply, and there on the dike's roadbed, I left his lifeless body for the eagle and drove on, caught up in my own past here.

A dozen miles down Highway 90, Eagle Lake Rod and Gun was always more than just a shooting club. For generations it had been a gunner's Augusta, and being invited to hunt there was a matter of deep symbolism. To shoot well at Eagle Lake was a stride toward genteel manhood, a way to become one of the steady adolescents our fathers had, since birth, looked forward to as sporting companions. So it was with awe that as an initiate I'd climbed the steep stairs into the lodge's rough-sawn bunk rooms, where important men sat with their black Labradors, swabbing the long blue barrels of their goose guns.

Then came early-morning coffee and biscuits brought by careful black men — not old, but obliged by the conventions of that time to assume the humility of age in deference to the club's hunting members. Outside, I'd slipped #4 shells into the cartridge loops of an old shooting coat, handed down like a vestment, in the cold black morning chorused with the surrounding voices of thousands of waking geese.

But out in the blinds, with their wavering strings coming in to the decoy spreads, I had refused to shoot. Just didn't put up my gun and pull the trigger. That was taken for cowardice. A fear of firearms, maybe, or blood-squeamishness, and afterward, though I was treated kindly, it was understood I would not return to Eagle Lake. Nor, by extension, continue in the society of hunting men, and I was eventually left to wander the same fields the hunters used, alone, with field glasses and a backpack sleeping bag I could roll out under any stand of cover. By high school, I knew I'd never return to that other world — the realm of hunting, sports, and business. It was the world that, with the geese and cranes I had always yearned to escape — and by then I knew I'd kept the best. I had not killed the big white birds I loved.

23.

Waiting for Nanook

As the calls of the Deadhorse geese faded, the downshift roar of a Cummins diesel and the crunch of the truck's shoulder-high tires on the gravel sent me floundering off the pipeline's elevated causeway. From its bordering ditch it was 10 feet up to where the driver already had his window down.

"Get on in. 'Fore you freeze."

I spread my hands at the gray landscape.

"It's forty, fifty degrees."

" 'Gainst reg'lations, walking this access road. People die out here."

"In winter maybe."

"Okee, Bub," the driver said. "But I'm keepin' my job. In about exactly thirteen seconds I'm radioing for Security."

He watched to see my reaction, which was to climb up and get on in, though I hated it. At the giant metal barns of the Prudhoe oil camps our flatbed eighteen-wheeler drove into an open airlock, waited for tall steel panels to close, then passed through an identical set of inner doors into a vast welding hangar. Beyond, through a maze of sealed bunk rooms I eventually found the building's center, which was its equally windowless cafeteria and lounge.

Like some sort of arctic space terminal, it was a place that, after life on the river, was unbearably claustrophobic, and the next morning I grabbed the first plane out, even though it was going in exactly the wrong direction: farther north, to Barrow.

From its airport I called Jen. As I knew they would, things went badly. Staying on in Alaska — who knew where — after my tours was what she'd feared I might do. And now I had.

Her anger was cold, but in her clinical way, it was also sad and reasoned.

"I thought I could help. Get you to grow up — take away all that hostility you've been expressing toward me."

All I was doing, I said, was trying to find Amelia.

"Alan. Whatever possibilities — any special things — that happen for you, I'm not going to be part of them."

I pointed out that, last spring, she was dying to follow falcons in the plane. She didn't answer, so I dropped that tack.

"You know — you have to know — that

in some way I must aggravate you."

Helpless, I looked at the receiver. But there was no appeal to Jennifer's psychology, even though I told her I was through in Alaska. For good. All I could do was listen as she told me that if I ever came back to Texas I ought to think about moving my stuff back to my old place.

I took a breath, but she was off the line. For an hour I panicked because Barrow's Post-Rogers Airport had only one flight out a day, and the next several days' were full. Then I settled down: there was a reason not to hurry — a polar bear had come in from the sea. The radio said it was prowling near the edge of town, where I had to go anyway, on Vose's behalf, to find the Wiley Post/Will Rogers monuments.

Barrow's depot of rusting, carport-size storage containers reminded me of George's sheet-metal mobile home, where, among the few interior decorations were a couple of big framed photographs. One was of a tall aviator standing under the wing of his plane.

"That's not me," George had said when he saw me looking. "People think it is, but it's not: it's Lindbergh."

He rubbed his thin hair. "Mighta been me: they called him Slim, and that's what they called me, too."

I told him I could see the resemblance, which actually was striking.

"Lindy'd gone way down in the public's eye by the time I started flying — he was against the Second War, you know. And his picture — one hung in every schoolroom in Maine, 'til it got taken down and thrown out."

George wiped the glass.

"Didn't matter to me. I got one of the negatives — cost me $75 to have it blown up and framed. And that was the Depression."

Next to Lindbergh was a photograph that didn't look like much: a pair of rudimentary rock cairns, a few yards apart. A scrap of text read:

THE CRASH HEARD 'ROUND THE WORLD

This monument is located along the shoreline of the Arctic Ocean, nine miles from Barrow, Alaska. It marks the location where Will Rogers and Wiley Post perished on August 15, 1935. Because of fog in the area they landed their Lockheed Vega on a lagoon just south of Barrow. They asked a group of Eskimos for directions, re-boarded their airplane and took off. Their aircraft lost power and nosed over, tearing off a wing and crashing upside down into shallow water.

George snorted. "That Orion — it wasn't a Vega — its engine only lost power because both its tanks were dry. I knew the guy wrote up the report. Post had been lost in fog all day and only landed 'cause they were out of fuel. Taking off again was just bad judgment: soon as he got the nose up, what little gas they had ran to the back of their wing tanks and stalled the engine. But writers always make it the plane's fault."

I straightened the photo, which hung in a perpetual left bank.

"Guys at one of the Barrow cafés still have some of that Orion," George went on. "But that doesn't mean they know their way around. They only come in for the summer, and the time I was there they steered me exactly wrong. Said the monuments were north of town. So I never found them."

He settled back in his green-checked easy chair, flushing a black cat that after one wild-eyed look at me streaked for the opening Vose had sawed in one of his trailer's doors.

"Peso," said George. "One-man cat. You ever get up that way — Barrow, I mean — ought to find those cairns. Tell me what they — that site, and all, is really like. You might even enjoy going out there, on account of the bears. No trees for a thousand miles — not that I could climb one anyway — but they been known to rip a door right off a rent car."

Even south of Barrow the monuments were hard to reach; it took a fat-tired three-wheeler to negotiate the black sand beach that was the only way to get to the little inlet where they sat, a few yards apart. Post's was missing its spired top, and a sign said the stone and mortar pedestals rested on the coldest spot in the United States.

"A god-forsaken desert of stone," in the words of Danish explorer Alwin Pedersen, Barrow's coast is, along with a few mountaintops, the most frigidly harsh of all the continent's biotic communities; that day, with a wind like frozen fire, it felt close to the yearly average of 19 degrees, and since nothing botanical ventures more than a few inches up into that gale, the dark, iron-oxide soil is bare except for lichen. The lake where Rogers and Post died was edged with ice, but on its frost-rimmed beach a scrim of red knots, their salmon breeding plumage now mottled into the gray of winter, hung together on the far bank. Any day they would leap into the air and swing away to the west, around Alaska's Bering shores. Then, pushed by northwest winds, for the next two months, always together, they would skitter from roost to feeding stop down the whole Pacific coast of North, Central, and South America, pulling up only where the land runs out 7,000 miles away at the tip of Tierra del Fuego.

Over the monuments, a scattering of snow buntings rose and dipped in frosty-winged waves. No polar bears, I'd have to tell George; the only one in the area was back near town. But the following night, as Barrow radio once more took up the admonition to stay bolted in, I slipped on my parka and stepped out into the darkness and first light snow of the season.

Like other northern ursines, ancestral polar bears were probably brown. Then, from the northern edge of the continent they went to sea. Far out across the ice of winter *Ursus maritimus*, the ocean bear, became a different creature — in ways more great cat than lumbering bear — killing big marine mammals with a meat-eater's explosive charge and changing his coat to ghostly white to live above the frozen sea.

Now, only in snow and frost, and usually in darkness (the ice bear sleeps much of the bright summer), does Nanook, as the Inupiat call him, truly exist. Even when that cold and dark incorporate the dim shapes of Barrow's outlying buildings, he is still the nightmare predator of the Pleistocene, supreme stalker of those long ages when the northern chill flowed down across the continents, sheathing the land in miles-thick ice. Nanook continues to move to the rhythm of that earlier day: with spectral deliberation he drifts over his jagged terrain, seeming almost sluggish until he strikes, erupting over 50 feet of floating ice to seize the head of a surfacing seal. Then, its face clamped in his great jaws, he may rip the seal's whole 200 pounds right out through its narrow breathing hole, crushing every bone in its solidly muscled body.

Yellow-white — not as the snow in sun, but pale as the shadows of dim winter snow — Nanook hides in flurries of wind-borne ice. There, says Inupiat myth, drifting snow squalls sometimes create a bear — ivory-fanged, black-clawed, and terrible — from nothing. It is the nightmare of every ringed seal breathing up into the darkness of winter, the terror of every isolated walrus pup and beached beluga whale. Huddled in the same dimness, I was not waiting for button-eyed cuddle bears, the mother-and-babes families of the postcards, for here lived a different Nanook, the pale, shapeless terror of dreams. Suspended in the drifting sheets of Barrow's first autumn snow, this bear could be anywhere. Or nowhere, dissolved into the wind-shaped swirls that bore only the idea of an ice bear.

If this creature should show itself, went the Zen, it meant perhaps that I was ready to see it. If not, I was not yet prepared to confront the nightmare specter of my psyche. Then I might know only that from somewhere in the mist its frozen eyes had instead seen me. And that should be enough.

The ice bear had already told me why I was here, searching the arctic night for a dangerous wraith that had no relevance to peregrines. It was why the cranes and geese had meant so much. And Amelia: why I'd looked for her so many miles and weeks after she was truly gone. Because all of it led back. Back to the place where for ancestral eons we were all small creatures living in the shadow of great carnivores — a place where, now, I'd put myself again, out past Barrow's last houses.

I had started back, across a patch of open ground when two dark shapes came hurtling past. They were not bears; they were teenage boys, and they were running for their lives.

"Bear! Fucking bear!"

The boys slid down an incline from the big dirt berm that holds back storm waves off the Chukchi Sea and, quick-dash distance from a lit doorway, spun around.

"Run, Dude! He's a big fucker!"

Skittering back and forth, both of them waved their arms, but in the darkness there was nothing to see.

"Where?" I called.

The taller kid dodged out from his doorway and sprinted in my direction.

"Comin' out now." He jabbed his hand to my right. "Over the ridge." Feet flailing, he kicked a flurry of dusted flakes as he reversed direction back to his door.

There was nothing to see, but I edged down Stevenson Road toward Nuvuk, the point north of town. And there the bear was. A yellow-gray smudge against the snow, he materialized from nothing, 100 feet away, moving fast over the rough ground. I stood like a post. Never seeming to touch the snow, in one smooth motion he cleared the earthen seawall and vaulted a ditch onto the paved strip where an early-season bowhead whale had just been butchered. Then, almost in the shadow of the Naval Research Station he paused and like a pale spirit reared up, swaying his flatiron forepaws left and right.

Far behind me, I heard the boys' door slam. Slowly, the bear dropped again and edged back onto the dim road. At a tire's icy rut he stretched out his neck — long, then longer, finally almost grotesquely long — to sniff its tread. That lengthy nuchal column, Canadian polar bear biologist Paul Watts later told me, gives big males dominance-establishing leverage in their wrestling bouts, most of which stop short of true fights because all-out battles have a temporary victor but almost always ultimately end with both bears dead of their wounds.

This wasn't one of those massive competitors. A young adult, he was little more than two thirds his potential weight of perhaps 1,500 pounds, but he was still twice as heavy as a big black bear. As I watched, he raised his head from the road and looked back over his shoulder. With eyesight better than mine, the bear could see me standing silhouetted and I felt a stab of fear. Undecided, he swung his brick-like head, scenting the air through wide-set nostrils, but all I could think of was the flat black holes that were his eyes. They were the eyes of white sharks and pit bull terriers, and they did not waver.

Neither did mine, for here was the wild I'd sought: before me in the darkness stood a huge, swift creature — as large a mammal as had ever hunted on land — with capacities infinitely beyond anything I could do to respond. If this bear chose to kill me, he would.

Then the silence faded, and as the moan of the wind under my parka hood intensified, Nanook dissolved. With no suggestion of movement, he ebbed into a buttermilk smear, then receded into nothing. As he disappeared, in the way that outside sounds worm their way into the fabric of a dream, the wind's cry gradually ascended into the shriek of a high-strung motor.

The boys had been on the phone, and as the screech of the snowmobile's engine drew closer I was bathed in the beam of a jiggling headlight. Above it, swiveling back and forth, was the face of a wild-haired, middle-aged Inupiat.

"Bear!" he yelled, pulling up beside me. "Where?"

I pointed across the tundra.

"Hop on!" he bawled, and the next second, as if shot from a cannon, we blasted out into the dark.

Within 100 yards the machine's track had left the sea, left the town, and was cutting virgin snow, layered just deep enough to let us slither sideways across icy patches and rocket over bumps of crusted willow. My terror forgotten — the bear itself forgotten — I found myself launched on Mr. Toad's Wild Ride. Mr. Toad combined with Extreme Motocross. Flopping up and down on the rear seat, I hung on to the machine's single unbroken grab-rail and, ahead, to a handful of Inupiat parka.

In the distance we glimpsed the bear, lit by the bouncing strobe of our light. He was scrambling up the side of a small gully, and to miss it we banked sharply right, where a stream's summer flow fingered into small tributaries. Our snow machine hit them almost flat out, porpoised into the air, then crashed down with a wobble before its spinning drive belt flung us up the next rise and into the air again.

It was phenomenal. As wildly kinetic as a roller coaster, our Ski-Doo cut across the edge of the bluff's drainage and dropped back onto the open tundra.

"Got him now!" Toad hollered, and I heard my own whooped exultation of agreement.

But we hadn't got him. Beyond our nose cowling, suddenly a still-unfrozen pond mirrored back our light.

As we slid to a stop I leaned forward. "What now?"

Revving his motor, Toad shrugged. The scent of strong drink wafted back over his beefy shoulders. "I'm Jim," he said. "Jim Skinner. Who'er you?" Before I could answer, his words were ripped out by the wind as we spun away from the water, arcing across the empty plain. Then as Jim looked back for directions I saw the end coming, and with a thunderous whump we plowed into the side of an unjumpable drift.

Stunned, for a minute Skinner and I sat in the dark. Then, his train of thought miraculously uninterrupted, he continued.

"Sometimes I'm Jimmy Lett, you know. Those are my Anglo names."

Jim punched the starter, but the motor just cranked, and as its battery drained, his shoulders sank. It was going to be a while, so he squirmed around on the seat.

"In First Nation, I'm Jimmy Nakoolak," he said in the soft Inupiat way. "That's my real name."

I couldn't have cared less, for it had come to me that, buried up to the thighs, we were helpless as snow-covied ptarmigan; now there was a real chance that Nakoolak and I could become the hunted.

"Jim. Bear wouldn't approach a snow machine? You know, the smell?"

He thought for a second.

"Don't really know. I just bought this thing."

But sitting waist-deep in the fallout from our snow explosion, Jim continued to consider the question.

"Out at the dump, you know, there's paint, diesel stink. And the bears, they eat everything — chew up car batteries for the grease on the terminals."

With that in mind, we pried our calves from the snow and struggled to haul Jim's Ski-Doo out of the drift. As we tipped it back onto its drive track, both of us puffing steam, he put a hand on my arm.

"Would you like me to spell Nakoolak? You could take it down."

We were halfway home when we met the

posse, a flotilla of screeching snow machines, rifles racked and ready since some of Nakoolak's First Nation pals had papers allowing them to kill polar bears for subsistence and native crafts. Jim hesitated, then agreed to take me on to town before he rejoined his buddies, but all the way in I was ashamed.

That bear wasn't going to harm me. His natural, and almost exclusive, prey was ringed seals. This was the sort of Nanook that always comes to town: a young male, counterpart of the boys who fled from him, then turned him in. A bear feeling his exploratory oats. The kind that people kill.

And what if he *had* attacked? Wasn't facing that possibility the reason I'd fled Prudhoe's stifling oil camps? Why I had slipped out here, into the nighttime streets, to taste the primal life that had defined the world for generations as recent as Nakoolak's parents?

"One day," goes a Native parable, "we will meet, bear of snow-dreams, out on the ice. And on that day it will not matter whether it is you that dies, or me."

That had been my vision, too. Before I'd betrayed it. On an off-road contrivance whose internal combustion power had, in seconds, triumphed over everything.

Bounding on it across the arctic plain had exposed my weakness — the same weakness that had let other yowling snow motors overcome the millennia it had taken Jimmy's careful people to accustom themselves to this quiet land. For I had been a willing, enthusiastic part of an attempt to run down my long-sought ice bear with a 40-horsepower motorized sled.

In minutes, that violent, euphoric machine had shown I was no different from the hermetically enclosed oil drillers of Prudhoe, no better than Eagle Lake's ceremonious gunners of geese. It had trumped my life out on the Colville; maybe even beaten Amelia. The days Vose and I had spent in the Cessna might have been only another mechanical conquest — our entry into the peregrines' world of sun and storm no more than a game of aviation thrills.

Maybe I had left Jen for . . . well, for nothing at all.

Part Three

THE BAY OF MEXICO

Wherever he goes . . . I will follow him. I will share the fear, and the exaltation, and the boredom, of the hunting life.

J. A. Baker,
The Peregrine

24.

Uniformes Y Documentarios

The flight out of Anchorage was packed with homeward-bound cannery and construction workers, and I was lucky to get a seat. Back in the tail end of Economy, I had a good view of the Canadian Rockies, but after jostling along in wind-bucking little planes all spring and summer, being able to gaze at hundreds of miles of curving horizon seemed like space travel. Yet as we slid down the eastern flank of the Continental Divide I was still caught up in the lives of falcons — some of them surely migrating southward below us that very morning. By now Amelia would also have left her young — this season's eyases I had wanted so much to see — to ride the upthrown currents past the same snowy peaks that George and I had paralleled, behind her, back in the spring.

It seemed impossible that my Colville young would ever see those mountains. Flight was a miracle just discovered in their wings, yet by mid-September they would have left the Arctic Slope and, without knowing they were doing so, set out. Below them would be some of the 120,000 Porcupine caribou, tramping south in broken lines toward wintering grounds in the Canadian Yukon. Soon their long columns would disappear in swirling storms and freezing fog, and driven by the same northerly winds, every juvenile peregrine would move farther each day in the same direction.

It was thought that most, probably still being fed by their parents, would wind through the passes of the Brooks Range, then perhaps penetrate the valley between the Endicott and Philip Smith Mountains. According to Alaska Fish and Game biologist Peter Bente, some might jog east down the Tanana or Chandalar drainages to the Yukon, after which there would be the barrier of the Selwyn and Mackenzie Ranges — north of the point where Amelia had cut west — to overcome before they reached the plains of Alberta and Saskatchewan.

But those were just guesses. Nobody knew where young arctic peregrines went — or for how long, if at all, they traveled with their parents. By the time a scattering of sub-adult *tundrius* came past the raptor-trapping stations in the southern Canadian and northern U.S. Rockies, the young seemed to be hunting for themselves. How many of them then reached far-off Padre was unrecorded, yet we knew that every year many of the juveniles made it at least that far.

Then, behind the liner's double-paned glass, it hit me. The weeks George and I had traveled with Amelia need not have been a waste. If nothing else, even with our mistakes we had learned a lot. About how to fly with peregrines, what they would do in different kinds of wind and varying weather, how relentless was the drive that forced them on. We had followed an Alaskan tundra falcon's northwest passage most of the way to her arctic home; perhaps now we could add a bit to the little that was known about where these falcons went, south of the Rio Grande.

Riddle had always believed that a substantial percentage of hatch-year *tundrius* made it as far as the Texas coast. But after that, he and others believed, as many as two-thirds of the juveniles born in any given year might die somewhere in Latin America. No one knew why, but if I could get a radio onto one of those young migrants, its southern route might lead us to some of the sources of peregrine mortality.

Vose agreed when I called him from Austin with the idea. Then he was silent, and remembering how difficult George had told me it was to get aviation repair done in Mexico, I could picture him reconsidering. Finally, he said that, falcons or not, he wasn't sure he really wanted to spend any more time in Latin America after all.

Even the mention of those potential plans would have set off another confrontation with Jennifer, so I picked up my truck and headed straight downtown to the law office of Burrell D. Johnston. An old Latin America hand, Burrell might be able to get the papers Vose and I would need to radio-track a peregrine into Mexico; my own inquiry had drawn a stern State Department letter refusing us permission.

Yet, as I gazed down at my North Slope boots slouched on his waiting-room carpet, it didn't seem like much of a shot.

"You may come with me now," announced his receptionist, leading me down a dark-paneled hallway, into which suddenly burst Burrell Johnston: beefy, graying, but only a few years older than me.

"Alan!" he beamed. "Did you bring me

another Jeffery?" Jeffery was a six-foot-long western diamondback rattlesnake I'd caught for him that, after too many viper-lawyer cracks, Burrell had decided to keep in a big glass cage next to his clients' chairs.

"Some of our combat fliers are coming to Big Bend!" he roared. Like Vose, Burrell had flown fighters, including the first wave of F3H Demon jets. "You and old what's-his-name . . ."

I waved him off the subject. There wasn't much time, I said. I needed to know if he was still involved with Partners in Progress for Latin America. Burrell grinned. He'd just gotten back from a funds-delivering trip to Mexico City, and after I told him about Alaska, the State Department's telemetry refusal, and how some of Padre's young peregrines were certain to go on into Mexico, Burrell steered me into his office.

From a drawer he pulled a huge, ornate envelope. "Been saving this for years," he mused. "Had to be a use for it."

Big enough for a coffee-table photobook and beige, with a cream-colored border surrounding embossed inlays that took up most of its surface, it was an envelope among envelopes. More important, it bore the seal of the Office of the President of Mexico. From it, Johnston slid a framed matte-finish photograph of a balding man in a dark business suit. In a sweeping hand the inscription read:

TE DESEO LO MEJOR

BURRELL D. JOHNSTON
JOSE LOPEZ PORTILLO Y PACHECO
EL PRESIDENTE DE MEXICO

On his bookshelf, Burrell propped the picture against a couple of F3H models. Then with a bow he handed me the envelope.

"Your Diplomatic Packet, Señor. Now where's that letter from State?"

While he read I sat out front, at his secretary's desk. The Mexican Presidential Envelope was impressive, but I saw room for improvement. There were plenty of office supplies nearby, and I stuck on twin ribbon rosettes of Mexico's red, green, and white national colors. They seemed to add a diplomatic flair, so I went further. One of the firm's members had been a member of the state legislature, and his office unknowingly contributed enough gold-leaf Texas House of Representatives seals to line the envelope's entire perimeter.

"Letter's baloney," said Burrell, striding back into the room, reading glasses dangling from one little finger.

"Mostly it is, anyway. Frankly, it's a gray area whether these clowns have the authority to order you not to track your peregrine anywhere you please."

He paced the room, mentally building a case.

"I'll come up with some kind of papers. Nobody from State'll ever see them anyway. . . ."

Burrell halted, grinning.

"But you can't do the trip in those clothes."

I looked down at my work shirt and jeans. "What, then?"

"Uniforms! Everybody in Mexico's got a uniform. Generals, guys who sweep the streets; it's part of their pay."

The more I thought about it, the more it seemed like a good idea.

"I always did like a man in uniform," cooed Mae West to temperance officer Cary Grant just before she invited him to come up and see her sometime, and I thought George and I might be able to benefit from the same sartorial advantage in Latin America. But that brought up another question. What kind of uniform?

There was a long tradition of counterfeit military livery — a gamut that ran from mere flamboyancy, such as General George S. Patton's personalized field jacket, adorned with more and fancier gilt buttons than those of his commander Eisenhower, to outright fabrications, like the generic officer's outfit that enabled author Jerzy Kozinski to travel unimpeded around post–World War II Europe.

I knew Vose wouldn't go that far.

But there *was* temptation. Like opting for military costumes like those the three scalawags — Nixon, Haldeman, and Ehrlichman — once ordered for the White House guards. Intended to add pomp and circumstance to the visit of British prime minister Harold Wilson — whom they figured would expect such finery after his European tours of state — Nixon's personal police appeared beneath black plastic replicas of the British Royal Guard's giant bearskin shakos, wearing brass-buttoned tunics that bore an uncanny resemblance to the hand-painted regalia of little lead carpet-soldiers.

George and I needed something tamer, yet still imposing enough to present a formidable impression, and since in our case even partial authenticity was out of the question, Burrell and I at last settled on a middle ground. Texas Highway Patrol.

Our outfits were surprisingly easy to get. From Johnston's office I called an Austin law-enforcement supply house and simply ordered four uniforms: genuine chocolate Highway Patrol polyester pants and bone-white, snap-shoulder military-style shirts. Two sets for me, another two, extra large, for George.

I worried about being asked for a photo ID, or even a badge. But when I went to pick up our uniforms nothing could have been further from the young clerk's mind.

"Yes sir, Captain Johnston," she nodded, hauling out a big box. "Your order, with the special monograms . . ."

I held up the shirt and pants.

"Plus the other set, the extra large?"

She nodded, then paused coyly.

"We sell to a lot of you officers . . . but, how come y'all got FALCON TRACKING TEAM on your shoulder patches?"

I couldn't help it. I leaned over and whispered our secret. My extra-large partner and I were going into Mexico on a Mission. She opened her mouth to exclaim, but I shushed her with a cautionary wave.

Still puzzled, holding thumb and fore-finger half an inch apart, she shook her head. "But . . . what kind of little tracks do falcons make?"

25.

Hurricane

To go with our uniforms, I had PROFES-
SIONAL BIOLOGIST business cards printed in
Spanish, and George got his neighbor, Dr.
Dennie Miller, director of the real
Chihuahuan Desert Research Institute — to
which we promised to someday donate the
radio-tracking gear purchased in its name —
to write a letter informing whomever it
might concern that Vose and I were re-
nowned bird scientists who should be provided with assistance and all due respect.

As I crossed the causeway over the
Laguna Madre toward Padre Island, I ran
our recharged scanner through its whole
frequency spectrum, including Amelia's
#.759. It was foolish, but I couldn't help
but hope that maybe she wasn't lost forever. If she was still alive I was sure she'd
come back down the barrier islands in the
fall, and though her transmitter's battery
had almost certainly gone dead, there was
a small possibility it could still be func-

tioning. One of the first test radios I'd set out on a fence post for Vose to locate with the plane had gotten lost, and I'd found it, more than four months later, only by driving down the same county road with our receiver set on SCAN.

Even from the bridge's height there was nothing but static, though for the moment that was fine with me. I was glad simply to be back on the island — which was possible only because a new researcher, biochemist Donald Morizot, had taken over the Cancer Center's peregrine study, leaving little or no Army presence on Padre.

Out on the beach, I hadn't yet gotten the Honda into top gear when it was clear that in late September the barrier islands were different. Because it was drier the wind-tidal plain was easier to travel, although there were far fewer falcons: with the Tropics just beyond, most adult peregrines went right by the coastal islands. Along the ocean, not far beyond the surf, clouds of laughing gulls swirled behind a pair of in-shore trawlers, and royal terns dipped over the breakers or sat along the beach next to dunlins and ruddy turnstones like the ones — maybe the *very* ones — I'd seen in the Arctic. It seemed like a feast for falcons, but the heavy vegetation of the nearby dunes offered too much cover for their prey, so a mile or two before Mansfield Channel I once more turned out onto the flats.

In the distance, across the honey-colored sand I could see part of an old drift fence. Its cedar posts were whitewashed with the droppings of gulls and egrets, and I idled the Honda forward, thinking how strange it was, after sitting almost in the nests of tundra peregrines, to again find falcons so wary. Yet, unlike sedentary city peregrines, accustomed to benign human crowds beneath their high-rise aeries, thousands of miles from their untouched polar nesting grounds shyness was all that kept the arctic *tundrius* alive.

An eighth of a mile from the fence I stopped and climbed up on my machine's cargo box. Nine feet in the air, my field glasses swept the skyline, picking out the dots that were the tips of the fence's weathered posts. One of them looked lumpy, so I eased toward it.

By the time I could make that lump out from down on the bike's seat, it had elongated into a falcon's back, bent low over prey. An adult tiercel, he was so small that at first I didn't take him for a peregrine at all. But with his feeding he didn't seem anxious to fly, so I crept closer. A hundred feet away, through my lenses I could see the silvery tufts of down he bent to pull from between his toes. A minuscule blue-gray gnatcatcher, it would never have attracted the attention of a female peregrine, and for even a willowy tiercel its food value would barely have repaid the energy it took to capture it. After thirty seconds the tiercel straightened up, glanced over his shoulder, and, seeing me, ducked his head in surprise.

Everything grew still. I breathed as shallowly as I could, but he calmly spread his slender wings, held them bowed against the air, then took off, casually folding the legs-and-wings remnant of the gnatcatcher up under his tail. A hundred feet beyond the drift fence he swung one foot forward, brought the scrap of a bird up for a final nip, then dropped it into the wind and headed across the bay.

Riddle's trappers had told me that scattered individuals like this lone male were likely to be the only adult peregrines I'd see on Padre in the fall, since a majority of the swift brown shapes that flitted in over the flats would be youngsters, gaunt from thousands of miles on the wing.

I couldn't wait to see them.

* * *

It was late afternoon when I reached the cabin, but the place was empty. Most of the trapping gear was gone too, which was strange in the middle of banding season, but all I really thought about was how my vanished tiercel had again ignited the fire. Telonics' phone number was still in my wallet, and when I called, the technician — an hour earlier on Mountain Time — couldn't have been friendlier. He remembered the Chihuahuan Desert Research Institute: we were a good customer, and he wanted to sell us more equipment, especially his new generation of 040 transmitters, smaller and lighter than the ones the Army had used, with delayed-pulse timing that extended the life of their batteries and produced an 80-mile line-of-sight signal. With that sort of relay, I told him, my partner and I would never have lost our peregrines last spring, and he said the new radios would work fine with our old-model receiver.

After Alaska, there wasn't much credit left on my Visa card, and it took every bit of what there was to get three of those fancy new transmitters onto the next day's FedEx. As we hung up, I noticed the answering machine's light and pressed the button. It was a message for me, from Jennifer. Out in the Gulf a hurricane was building; everyone was supposed to evacuate the barrier islands.

No wonder the guys left, but I wasn't sure why Jen had called. "I thought you were through with me," I said when I reached her. Yet it was as though nothing had happened: Jen knew I was nuts about being in storms and wouldn't leave the island. So she was coming down to get me.

I told her not to bring her little Nissan; it was too light. "Get a junk rental: big old sedan. I'll meet you, and we'll go back out to the beach to watch the storm come in."

I had to listen awhile, but finally I was able to say, "I know, it's a hurricane. That's what's so great about it."

As the storm moved slowly toward shore it hooked north, so I drove up the coast, aiming for its new landfall. By the time I rendezvoused with Jen at El Campo there was still no wind, but the sky was the wonderful cobalt green it goes to right before big weather, and in the dank air that was sucking in the tropical depression we loaded the four-door Ford she'd gotten from Rent a Heap Cheap. As darkness fell, we slipped — the only car on the road — onto the series of low bridges leading out to Matagorda Island. Years of chasing hurricanes had taught me that you had to get in early, before the police barriers went up.

Then, ahead, I saw smudge pots.

It was only a local deputy, who was hauling out the equipment for that evening's evacuation and hadn't gotten his final instructions. He still wouldn't have let us by, except that I showed him my *Texas Monthly* identification and Jennifer told him that I'd covered hurricanes for the magazine before.

Then, with further luck, after finding everything boarded up on the island, we managed to get a room from an elderly couple who were closing their beachfront motel to go inland. They were glad, they said hurriedly, to have someone staying over whom they might be able to call after the storm to find out if their place was still standing. I told them everything would be fine, but that we'd let them know anyway.

During the night the storm became a hurricane, and as its counterclockwise rotation reached land the wind swung around to the northeast, battering the building's red-brick walls with pellets of rain. In the morning, we could see from our big window that foaming breakers were running up the beach almost as far as the rooms below, and when I opened our door the wind punched me back a step. Jen ran to help me close it.

That brought her face next to mine, and her soft gray eyes were a little scared.

"We're leaving," she said.

"But it's wonderful. Just getting wonderful. All this wind."

"It's too much. This is dangerous."

She started to put things into her case.

"Wait wait wait. Let's go out, just for a minute. Then we'll zoom back over to the mainland before the real storm starts."

Jen studied my expression, but I held up a solemn hand.

"Promise."

The rain had stopped, yet the sky was darker than in any rain, and even going down the steps Jen and I had to hold each other to keep from being blown over. Out on the sand, I could lean so far into the gale that, reaching down, I could trail my fingers in the grains boiling in flat swirls across its surface, then gusting up to stick to our bare legs.

With every minute the wind grew louder, but it didn't roar. Filled with scraps of paper, green plastic bottles, and chunks of Styrofoam, little by little it swelled into a universe of sound that pounded every cubic inch of air, abrading not just our hearing but every tactile sense.

It was a day for falcons, I yelled to Jennifer, looking hard every time a set of big frenzied wings flashed by. But we spotted only disoriented pelicans, cormorants, egrets, and a huge, desperately flapping great blue heron. On the beach, the combers were still growing, then collapsing forward with a concussive boom we could feel through our soles as their impacts flung ribbons of creamy foam into the wind. We skirted the edge of a newly dug bay the waves had cut from the sea, and I realized that everything was perfect: it was a storm right on the brink of overwhelming our ability to live in it.

Jennifer could also see that almost-lost control.

"Alan, you are getting too wild," she yelled, fingers dug into my arm. Like a hunting hound, I kept tugging into the wind, but she grabbed my face and forced it toward her.

"We . . . are . . . going back. *Right now.*"

On the opposite heading, pushed by the wind, we edged along the landward fringe of the ever-narrower beach, sidestepping the waves that by now ran almost all the way to the dunes. Then Jen pointed: the expanding inlet we'd waded on the way out had become a channel, slicing deep into the sea oats. To the west a similar incursion was spreading oceanward from the laguna.

In between was Matagorda's ridge of grassy dunes. It was our only choice, and we ran and tripped and ran again, up into its tangled Gulf paspalum. Below the largest dune a hollow was matted with blown-down grass and goat's-foot morning glory, and together Jen and I rolled down into its recess.

I was still gasping from the wind when I realized Jen was shaking with laughter. She raised her head from my chest and, fighting the gale, leaned around to my ear.

"This is really, you know, why I came. If I can just keep you from killing us. . . ."

She looked out at the chaos swirling past our shallow refuge. Unable to hold their grip on the sand, black-and-white-winged willets, oystercatchers and stilts, even strong-winged royal terns spun by, thrashing like feathery foliage. Jen reached out toward them, then shrieked as her arm was blasted downwind. I looked at her in wonder: twenty minutes before she had been trying to drag me off the island.

"You know my dad," she shouted.

"Dumpy little dentist. What would he do here?"

"Have a stroke," I said, "probably."

Jen clambered over onto me. "Then let's give him something." She buried her face in my neck. "Something to have a stroke over."

It was afternoon when we finally saw the motel, barely visible even from the beach. Waves had knocked in some of its air conditioners and flooded all the lower-level rooms, but the top floor looked OK, and shedding streams of water, we struggled up the steps to its balcony. One of our windows was cracked, but it was dry inside as we collapsed on the rope-rugged floor. We lay on our backs, breathing hard, staring at the ceiling it was not light enough to see. After a while we turned and, even in the shadows, caught each other's eyes. I eased a mat of Jen's russet hair away from her face, and slowly she pushed back the thinner strands from my own forehead.

We were quiet a long time.

"You're not really going to follow these falcons?" she whispered. "Not into Mexico . . . ?" and I shook my head no.

But after the storm I went back to Padre. I told Jen what I'd come to believe: that even though there was almost no chance of finding Amelia on the Texas coast, George and I — even Jennifer, if she wanted — could maybe hook up with one of the adolescent peregrines that were sure to come through later. Just like last spring, we could follow it up and down the barrier islands. Then, at most, a little way over the Rio Grande, because Vose had made it clear he wasn't going one bit farther. Jen said she assumed that I could see the sense of that. Then, to my surprise, she dropped the subject.

Down at the south end of Padre the storm hadn't pounded the coast so hard, and there were hundreds of wading birds still on the flats. No peregrines, but with regular battery charges I could keep the scanner on all day, listening for Amelia and C.L., though I knew there was almost no chance of finding either one. By staying on the move I used up all three of my extra gas cans, and the next night I went in for fuel.

There was a new bunch of trappers at the cabin, most of them recruits from Morizot's Bastrop research facility, and I was happy not to see a set of military fatigues among them. But telling everyone about my Alaskan fledglings made me feel

432

like a grizzled pro, especially when Don handed me a little brown box containing whatever might have been important enough to warrant an overnight FedEx.

Nearly another week had passed when, just south of Mansfield Channel, I made out a dot on the horizon. Last spring a speck like that would have been only an upright stump, but I now had something of a falcon's eye, and from the way the lump sat on its driftwood perch my visual template said peregrine. This one's slump suggested it was also a sub-adult, which meant that for days it would have been following the also-migrating shorebirds it was probably only rarely able to catch.

Two months before I'd watched youngsters like this fight one another on their polar nest scrapes, footing even their parents in the fierceness of their hunger, then stroking the air with all their might just to approach a fleeing prey bird. To such a ravenous youngster a tethered pigeon would be irresistible, so I rolled the ATV back far enough to drop below the falcon's horizon and threw a pigeon up. I knew its abrupt flurry would intercept the hawk's line of sight, and though I never saw the falcon clear the skyline, in a flare of reckless wings she was suddenly right over me.

Fearless, she hovered almost to a stop. From her brown-stippled breast tufts of down still sprouted, which meant I'd been right: this was a juvenile, ferocious with hunger, talons grasping for the pigeon as she dropped, heedless of me, of my ATV, and of the pigeon's transparent veil of monofilament loops.

I was using a heavier drag this season, and the young peregrine went only a few yards before her snared toes pulled her onto the sand. There she gazed up with angry, terrified eyes, and as I stooped she hissed and pulled away, wincing at the travesty of my touch. But I could move swiftly now, and, noting how light, almost skeletal, she felt, I stitched my new transmitter to the base of her central retrices. As the mask slipped off the pale ocher of her brow and cheeks — markings that told me she was a juvenile *tundrius* — she was so like the Colville nestlings I could hardly bear to let her go. There was little chance she would stay on Padre, a spit of sand she was only passing on the way to her distant destination, and no matter how far my transmitter might allow us to travel in each other's company, once she was out of my arms I knew I'd probably never see her again. But now she was my girl, and even as she lay in my hands I worried about her and where she would go.

In her short life she had seen chalky polar bluffs, swift-water rivers, then perhaps days of alien grass that swept up to the Rockies' eastern wall. From only that, how could she find her way farther? Locate, somewhere far down over the vast blue curve of the earth, a tropical coast she had never seen — a place where the sea glowed turquoise and the shoreline baked in a tropic sun?

For her destination, despite its mystery, was no dream and attempting to reach it was a quest as real and vital as Amelia's. Within minutes this young creature would again be embarked upon the age-old journey of all her kind, and gradually her shiver became my own. Nothing I would ever do, or know, would be as important as this newly flighted falcon's blind pilgrimage of faith, and in a flash I knew that I would go with her. I dropped the dead pigeon onto the sand, hoping she'd come back later to feed on it, and set her into the air, where she spread her wings and rose like an airborne spirit, winnowing away on the wind.

26.

Anukiat

Down at the end of the Del Rio's parking lot, collapsed on its failed shocks, was Cameron Field's old brown Chevy. My honk brought Vose out, arms filled with freshly bought supplies — groceries he figured we'd need for even a few days south of the border, where he didn't trust the food. Plus, he'd had a lot to do at the ranch, where his adobe walls, despite the polyurethane, had deteriorated from summer rain; when I called he was still indignant about the group of archaeologists who'd hiked up out of one of his canyons and pressed him for details about whether the waist-high dirt rectangles they'd spotted on an aerial photo were of Native American or pioneer origin.

"Originated outa my own two hands," Vose said he'd told them. "Close to twenty years' labor."

In a better mood now, he stowed the groceries and the Telonics receiver, our three new transmitter frequencies taped to

436

its leather case, in the Chevy's backseat. Then, recalling the *mordidas* — shakedown bribes Vose had said we might face at Mexican airports — I pulled out my shiny aluminum briefcase and thrust Burrell's Diplomatic Packet at him.

It was sealed with a gob of red wax, so George didn't try to open it, just turned it over and over in his big freckled hands.

"Credentials," I said.

He nodded, thumbing the packet's gold seals and many-ribboned rosettes.

"Looks like a flat piñata," he finally observed. "Don't know if that's good or bad."

I took it to be good, and shook George's hand — for a long time, because we hadn't seen each other since spring. Then we agreed that, while I tried to get our remaining two transmitters onto peregrines he'd keep track of our new gal. I tried to describe her, but all I could remember was the shudder of her bony breast. Her fragility. And above all, how alone she was. Not like the geese with their gypsy families trailing across the country — by this point on her journey south, my youngster was as alone as it is possible for a creature to be.

And she was in jeopardy. Like almost every arctic adolescent, even as a bringer of death to other birds she was barely up to her task. Her slender shanks, still a youthful chartreuse, had strength, but not brute power. Those genteel toes could slip a razor-pointed talon into tender flesh, but without the momentum of a dive — a maneuver every maturing peregrine takes months to learn — her claws might not kill a bird of even her own size.

Her wings were strong too, but equally delicate. Long and sleek, with rounded tops and concave undersides, they weighed almost nothing: together, 7 or 8 ounces of flesh and feather. They reminded me of George's first plane, a Super Cub Piper, whose entire wing a man could lift and carry by himself. That wing, too, was joined with glue — the animal protein keratin, which had stiffened the cloth covering the Wrights' Kitty Hawk Flyer and was the shaping matrix of every glistening flight feather that made up our youngster's airfoil blades.

Yet even with this adolescent's inherited potential for transcendent skill aloft, just to continue her migration would be a struggle, I told George. Thin as no adult peregrine ever is, this youngster would have no leeway for months, no reserve of fat to tide her through a week without prey, or past even a slight injury. Hovering above me, she had screamed — in rage I'd assumed, though perhaps in triumph that she was at last about to capture prey; even as far as she had flown, my pigeon could have been among her earliest kills.

That was because on her still-developing wings she could outfly fleeing birds only at the very top of her game. Greater aerial skills would come later, if she lived, giving her the chance to survive in terrain where prey was scarce or hard to hunt, but at this stage in her life anything less than physical perfection and she was dead.

Vose said that meant I'd better get our remaining two transmitters on a couple of other falcons pronto, just in case, but as he headed for the plane, I knew that George, too, had already begun to feel that now this skinny girl was our peregrine, her life, with all its difficulties, had become our own. Maybe Jennifer would also understand.

By the time I'd unloaded my truck at the end of Padre's National Seashore road it was late enough for peregrines to be on the wing, but the rest of the way up the island on the ATV I spotted only gulls, cormorants, and willets. After dark, sleeping bag tucked against the Honda's fat rear tires, I could hear above me the rustle of my pigeons, cooing softly in their cantilevered coop. With the three of us huddled there in shelter from the wind, I realized I'd begun to think of my decoy birds as companions. I had always released every one that survived a single falcon-baiting session, but having grown more empathetic to peregrines, I'd also begun to feel the same toward their prey, and hearing these two low through the night gave me the shivers.

"OK, OK," I muttered up at their box. "Tomorrow you're both getting let go."

But it wasn't so easy. There hadn't been a falcon visible on the flats all day, so I thought the pigeons had it made when I turned them out at dusk, a gray street mongrel and a big sleek homer. The homer — a pure white cock able to fly nearly as far in a day as Amelia — shot off westward toward the laguna. Beyond, the mainland was an easy 15-mile hop.

Or it should have been, except that I had forgotten. Peregrines respond to nothing more strongly than to the flash of alabaster wings, and four or five stories up, my homer was flat out when a dark scimitar slid in behind him. A falcon, resting out of my sight over the horizon, had swung into the air.

I spun the Honda and took off after

them. Certain of the pigeon's vulnerability, the peregrine — a juvenile, big but seemingly unsure of how to handle a bird the size of this homer — tilted back and forth above her prey, for she was fast and strong and the homer was not going to outfly her. Still, he had maneuvers: when the falcon feinted down the homer clamped into a tight projectile and rolled to his right, using up half his altitude but putting an additional 50 feet between him and his pursuer.

That panicked flash of white feathers was all it took to rivet the eye of every falcon on the flats, and immediately another thin shape came raking in. Then a third joined them, and in a maelstrom of slender wings, four young tundra peregrines tumbled along behind the homer, each barring its competitors' way by twisting onto its back with outstretched talons.

Forced down almost onto the sand, the homer swerved back and forth in desperate arcs that let me close on the group. I wasn't trying to trap, just to intercede, to save my hapless homer from the storm of wings and claws into which I'd delivered him. Eighty yards back I had the Honda pegged, but I was too late. The first big falcon slipped below the interceptive swoops of her rivals and in a blink smothered the pigeon within her wings. My momentum carried me almost onto the patch of algae where they fell before I spotted them — the peregrine's huge eyes aflame, her spread remiges mantled over the plump white homer, who stared out even wider-eyed from beneath her breast.

I was sure that, as a juvenile, this falcon had never tried to lift anything as heavy as that big pigeon, yet, mouth agape and panting, she was so bound to her hard-won prey that in a frenzy of flapping wings she tried to drag him across the seaweed. As I bailed off the saddle, the falcon loosed her claws to pull away, but before her wings could grasp the air I was on top of them. My dive knocked her onto her back, and before she could right herself I held the tip of one long wing.

That was a bad idea. She could easily have jerked free, leaving me with a handful of primary feathers and her with marginally functional flight. Instead, she flailed around to face me, sinking all eight talons into my forearm, where I could reach them with my left hand.

Even moving as quickly as I could, by

the time I had the big falcon hooded and body-stockinged it was pitch-dark. That meant it was better to take her back to the cabin to be fed, then fitted with my transmitter in good light the next morning. But as I lifted her it was obvious that she needed no fattening. Her chest was packed with muscle, and I realized that it was not by chance she had outflown her competitors; already, this youngster had learned to kill.

Meanwhile, my right hand, into which she had sunk all four free claws, had begun to throb, and after ruffling my poor pigeon's plumage and finding no wounds, I eased him back into his coop and wrapped a bandage around my palm. Then, beneath the idling ATV I caught a glimpse of feathers. Recognizing its danger exposed on the flats, my second pigeon had refused to leave, sweeping along behind the Honda as I'd pursued her companion, then fluttering in between its rear wheels when I stopped.

Gathering her up, I put her in with the homer, and that evening drove all the way into town to release them on South Padre's pigeon-filled jetties. That meant that by the time I got back to the cabin all the trappers were in. David Williamson, a wiry master falconer and veteran of Riddle's program, had also brought in a hatch-year female similar to my husky northern juvenile. A sleek, already banded second-year male was also at the cabin.

We were about to unhood, de-stocking, and feed all three captives when project leader Don Morizot walked in. Ordinarily the best-humored guy in camp, a Vietnam survivor whose sardonic humor had gotten him through two years of jungle combat, Don took one look at our three captive peregrines and exploded.

"What the Hell are falcons doing in here?" he yelled. He waved both arms. If he were a wildlife warden every one of us would be under arrest — for a state as well as a federal offense.

"Don," I said. "Whoa. Riddle and Bill Satterfield brought in birds all the time. Took some back to be operated —"

"Ken Riddle is off in Saudi Arabia or some damn place!" shouted Morizot. "My U.S. Fish and Wildlife permit — which covers every one of you — says, and I mean states explicitly, that we are to hold peregrines only for the immediate duration of banding and the withdrawal of a blood sample."

He paused, puffing.

"So get 'em out of here."

I couldn't let my falcon go; her transmitter was still packed on the ATV.

"It's pitch out there," I said. "These birds can't fly."

His fair complexion flaming, Morizot looked hard at me.

"They are *going* to fly."

Don started for the bound peregrines, but I bundled all three up before he got there.

"I'll take 'em," I blurted. "Right now. Out on the deck."

Williamson, whose adolescent peregrine I was also carrying, came with me, but as we cleared the back door I told him the porch's floodlight might blind the falcons, so I'd walk them down the beach to where it was darker. To my relief he said OK, but be sure to bring back his hood.

Otherwise, I'd have been in trouble. Morizot had no permits for radio-tracking and would have really gone berserk if he had known what I was up to. But with David gone I had a few minutes, and from the Honda I grabbed a hand lantern and my last two transmitters. Behind the nearest dune I knelt, light wedged into the sand, then stitched and Super-Glued one of the slim new transmitters onto my plump adolescent. For seconds I gazed at her pale cheeks, gold as a harvest moon, and at the dark marbling of her amber flanks; then I tossed her up, hoping the wind was strong enough to carry her away from the spotlight glaring from the cabin's roof.

Time was running out, and I'd almost finished sewing transmitter #.973 onto the next peregrine when I noticed it wasn't Williamson's hatch-year female. It was the banded second-year tiercel. I hadn't thought of putting a transmitter on him, but any second, either Don or David could come looking for me, so I scrambled to the top of a dune and pulled off the little male's hood.

Instantly he became a tiger, hissing and fighting with strength beyond his size. So with everything I had I launched him — a small, muscular torpedo — high over the beach. In my lantern's beam I saw him go up, unfurl the spring-steel blades of his wings, and dig them into the dark air as he arced out over the breakers. Then, without another transmitter to put on, I set David's young female more gently into the air, watching again with the lantern to see that she also went south, on the wind, away from the cabin's light.

By morning all Don's bad will had vanished. Still in darkness, he and his trappers fueled their ATVs, but today I took no pigeons and left in my truck before the rest. I was through on Padre, and what I had to do now was reach George before sunup. From the first Exxon, I phoned to tell him to get '469 off the runway as soon as it was light, then run the scanner on all three frequencies. Afterward, I called back to the beach house to tell Morizot no hard feelings, but that I'd had to take off on a different project and wouldn't be there for the rest of banding season.

Instead of Don, I got one of the new guys, who said he'd pass the message on. Then he told me that, after hearing about my time in the Arctic he'd noticed in an old trapping log that one of the nestlings we had banded with Riddle on the Colville had come to Padre.

Chances were, it wasn't mine. I had only banded babies from two aeries. The first held a pair of females and an unhatched egg; the other was a brood of three big falcons and a scrawny little male. He was the only tiercel I had banded, and for me that had been a special nest. The day before we'd found it, I'd come across the silvery stump of a driftwood fir, waiting for some future flood to send it tumbling downriver

447

like Paddle-to-the-Sea. The fir's trunk was split with big cracks, and inside one I had noticed feathers.

Austrian raptor biologist Hans Frey once told me that in checking peregrine aeries in the Alps he'd occasionally seen such larders: crevices where a tiercel peregrine had hidden small kills as caches for his mate and offspring. From that Alaskan stump I'd dug a yellow wagtail, its missing head severed in unmistakable peregrine fashion, then the small bodies of two savannah sparrows — fleas of the Arctic, biologists called them — that I stashed in my pack to photograph as examples of cached prey.

I didn't know if this sort of stash had been reported on Alaska's Arctic Slope, but in this stump a peregrine had clearly stored its prey, and for hours, from a distance, I scrutinized the log. Neither parent slipped back to the pantry, and as we shoved off the next morning the tree was still unvisited.

Not far from that larder was the aerie I remembered, set on a tussocked slope and hidden from the river by a ruff of green bunchgrass. I'd thought it might hold newborns, but as I peeked over the hollow's uphill edge I saw the eyases were a hefty three weeks.

Old enough to have seen me coming and not liked it.

Swathed in dirty wool, the whole brood was reared back on the stumps of their emerging tail feathers, hissing with open beaks, so I moved back to not frighten them further. I'd seen enough half-grown peregrines to recognize them as females, so squatty-fat that except for their hooked beaks and ferocious feet they looked like steroid-bulked poultry. As I watched from a little way off, all three gradually forgot their terror and, eyelids drooping, leaned against each other to snooze. Very slowly, I moved forward.

It wasn't slow enough, for the trio snapped up their heads and, as well as their sumo shapes allowed, scuttled to the far end of their grassy basin. That let me see their brother. Too puny to scuttle, and not much more than half his sisters' size, he'd been hidden beneath their doughy bodies. Now he just lay there, looking up with a baleful and — it seemed to me, given his circumstances — despairing expression.

From his slender frame and gracile skull I decided he was a tiercel and, because of his small size, sunken breast, and scrawny drumsticks, the last of the chicks to hatch.

"You all are gonna make it," I told the three big girls, then looked at their brother. "Not you, Little Guy. Probably."

I picked him up, curled in a defensive ball. He was just insurance, in case a late-season freeze killed his older sisters, since in these arctic aeries the last-laid egg may endure the cold to hatch later. Otherwise, he'd not be likely to survive competition with his larger nestmates. In most places, peregrine eggs — usually deposited two days apart — don't begin to develop until incubation starts, after the last egg is laid. But the Arctic Slope's fierce weather means that incubation may begin as soon as the first egg is laid, giving firstborns as much as a week's head start on younger siblings.

Settling the little tiercel in my lap to put on his U.S. Fish and Wildlife band, I eased one knotted fist out from his breast.

"You know those big girls are gonna try and kill you."

Looking over at his competition, the older sisters who would soon be threatening one another with their hooked talons, I put a band on his leathery shank.

"Stay alive and you'll fly first," I told him. "Guy peregrines always do."

As I set the tiercel down, he flopped

over, arms outstretched like a fuzzy lizard. Beyond him, the nearest female was ready with a swift swipe that almost caught my hand but let me grab her other leg and turn her around so her claws pointed away from my ribs.

Her two squirmy sisters came next.

"Now. You've all got lovely silver bracelets," I told them as the last one went back. "Very exclusive."

Out from under the cover of his sisters, the little male was shivering, and I thought of something. A biologically incorrect something: like Bill Satterfield nurturing cripple-legged C.L.

In my pack lay the still unphotographed prey-bird bodies I had snitched from their driftwood larder.

"Chup, chup. Pee-chup," I called.

The little guy rolled his too-heavy head and looked up. On both sides of his breastbone the skin was limp, and I knew my specimens had a better use than scientific data points.

Its flesh-colored toes brittle as twigs, one of the savannah sparrows was still moist inside, and I offered the tiercel half its breast. "It's OK; you'll like it. Pee-chup, chup."

He wasn't interested. But I lay both the

sparrow's twin and the now never-to-be-documented wagtail on the tiercel's side of the aerie.

"I know, I know. None of this is gonna do you much good. But I want you to live, hear?"

From South Padre, I called back to the trapping cabin, but everyone had gone. For minutes I harbored hopes that this year's banded tiercel might have been the little arctic guy I'd held, two seasons ago, as a downy chick clinging to life on a cliffside ledge just upriver from Booming Bluff. But my wishes could not outrun reality: there was, perhaps, one chance in a hundred it was him.

As I headed for the airfield, dawn broke over the high-arched causeway, lighting the channel's muddy verges, where rows of migrant shorebirds waited for the sun. Many of them had recently arrived from the Arctic, and as I imagined their long journey, suddenly I realized that the opportunity that still lay before George and me was itself almost too incredible to believe.

Some of those sandpipers and dowitchers could have raised their young within yards of the nest scrapes of the falcons I had seen. Probably had. No one would ever know which ones, but right now, back on Padre Island were three tundra peregrines who in all likelihood had just come down from aeries along the same Nearctic rivers. And one of them was not a puzzled adolescent on its first, wandering trip south, but a second-year male — part of the minority of northern adolescents who manage to return to their far-northern homes. Now, presumably knowing where he was headed, he was on his way south again, and Vose and I had a chance to go with him.

As I came off the bridge I had the gas pedal on the floor. George ought to be loaded up by now, fueled and ready to take off; if we could get behind any of these three young peregrines we might be able to discover a clue to why so many failed to return in the spring.

The reason might simply be storms, bad weather, scarce prey. But if it was something artificial — of human/chemical origin — then it was likely to be something that could be corrected.

Still, I couldn't stop thinking about that first Colville aerie. What had I called that scrawny, barely surviving tiercel? Little Guy. That was no fit name, even if he had died there by Booming Bluff. But if he'd somehow stayed alive, I thought, by now he could be a twin of the muscular male I'd launched out over the Gulf.

How about what that Inupiat kid had called me? Hoping I had drugs for him, through tobacco-stained teeth he'd complimented me on my travels, grinning something like *unugliat* or *anukliak*. It didn't matter: *peregrine* means "wanderer," anyway, and now — I told Vose when he stepped down from '469 after a successful reconnaissance an hour later — our last transmitter peregrine was Anukiat. Quasi-Inupiat for "Traveler."

27.

La Pesca

Neither of the two young females nor
Anukiat left South Padre that first morning,
nor did they for the rest of the week. That, I
hoped, would give Vose time to get attached
to them, though I knew better than to press
the issue. At first George had been reluctant
to follow Amelia, too. So, on many of our
tracking days I left him to do the radio-
telemetry alone, while I tried to arrange lo-
gistics for what I hoped would be a longer
journey than Vose had so far agreed to take.

On his solitary tracking days George was
always back by dark, after all three falcons
had gone to roost, and as I paced
Cameron's flight apron, waiting, I heard
approaching steps. It was Guzman, a Mexican pilot we had met the previous spring
but still knew only by his last name. As
usual, he wanted to talk transmitters.

Especially the new ones Vose had told
him I was bringing. Guzman owned an elderly Beechcraft twin that he used to haul

soft contraband — computer parts, electronics, video stuff, he said — from Cameron down to Tampico and Veracruz, and our tiny radios had given him an idea.

Hide those *pequeño* transmitters under the Federales's vehicles, he volunteered, arm around my shoulder. That way his guys would always know where their enemies were. I told Guzman he'd need a good receiver as well, and gave him the manufacturer's phone number. I also pointed out he would have to pose as some sort of wildlife researcher to deal with the company, but he said no problem, and that in return for the tip he'd bring George and me as many supplies as we needed in Mexico, though how he planned to find us remained a mystery.

Just by following our falcon radios, he laughed darkly, folding Telonics' number in his wallet.

Still no Vose, but while I consulted with Guzman, a bone-white Mitsubishi jet had swept onto the runway. Like a mini-airliner it whined down its turbines, and as its air-lock door whooshed open, two executives and their secretaries stepped down. They were tiptoeing gingerly across the crumbly asphalt when the four of them, plus Guzman and I, looked up. High wings set above its plump, orange-striped abdomen, overhead buzzed '469, cabin suspended from its stalk-like wing struts, long, triangular antennas branching from its right and left brow ridges like a huge metal dragonfly.

George wagged his wings and I knew he'd seen me, and that again he'd found our peregrines. Before he had the prop turned off I'd come in through the passenger door to hear the details. All solid pulses again today, but way down on Boca Chica Island, at the mouth of the Rio Grande.

That was where Mexico began, which gave us only until morning. So on the way back to the Del Rio I told Vose how worried Jen was about our going across the border. His answer was simple. Only be gone a couple of days: bring her along.

George had always had a soft spot for Jennifer, but I maintained that as long as we had a signal from either of the two young females or from Anukiat we couldn't let anything stand in the way of following our birds wherever they went.

Yet as I pulled into our customary Burger King it was clear that Vose didn't see it as quite so sacred a mission.

"You still thinkin' this bird might be

yours from Alaska?"

After our Canadian border crossing, I wasn't going to deceive him again, so I told George all I knew. Much as I wanted Anukiat to be that now-grown Colville nestling, it was extremely unlikely that he was.

Vose looked even more dubious and with a scowl opened his Whopper. At the very least, I went on hurriedly, with one of our long-range 040 transmitters on this adult tiercel we'd have a far better chance than with either of the hatch-year females that he'd make it all the way through Mexico and Central America. Maybe even on past Panama to the Southern Hemisphere, where I knew Vose had, in the past, at least mused about following a falcon.

George thought about my rationale and said nothing. Out at the car, as he lowered himself through the passenger door I saw he wasn't as strong as he'd been back in the spring. He saw me watching, but he wasn't angry. Slowly, he broke into a grin, and I saw the old long-distance aviator's glint flicker across his eyes.

"Mister," he said, "this last falcon — your band or not; from Alaska or Peoria, Illinois. He's headed somewhere. And that's as good an excuse as any to go after him."

He paused. "But Jennifer still ought to come with us."

The plan was to clear Customs early, but before our alarm went off, the phone rang in the darkness. Jen had gotten my message about Vose and me going south. "I'll bring my stuff," she said. "Be with you when that old thing gives out."

George overheard her.

"Our aircraft is not going to give out," he proclaimed. Then after a moment, "Or me either."

It would never work, I told her. All three falcons were already right at the border and in less than an hour we had to take off after them.

"It's my choice," she declared quietly. "My commitment. Dumb as this trip is."

"We could be gone weeks, Hon. Months. And you haven't got a passport. . . ."

George interrupted to say she could always leave with us and then just go home, but Jen didn't hear him.

Get that passport lined up, I told her, and you can fly down and join us. I said I'd call her along the way.

There was no answer. Jen didn't hang up, just put the phone down and went away. Finally, I was the one who clicked off.

★ ★ ★

It came as a surprise but the two peregrine girls went first. They covered only a few miles, not together, and they took all the next day to do it. What George and I hadn't expected was that as soon as both falcons crossed the Rio Grande they angled inland, past the northern tip of the little Tamaulipan mountains that lay right behind the beach.

By midday, we were still with them because, to our relief, the big U.S. Postal Service eagles that George had extracted from his mail carrier and glued to each of the Cessna's doors had attracted no negative attention at either U.S. Customs or its Mexican counterpart. Probably, I proudly pointed out, because of the imposing impression conferred by our new Highway Patrol uniforms, though Vose still wasn't happy with the airless polyester of his starched new ensemble. But as we lifted off over the river, it was with anything but elation: George's mother was gravely ill, as was my grandmother. And without Jen in the co-pilot's seat, I knew I'd burned my last bridge.

For once, though, the peregrines were easy. Our two adolescent girls just dawdled, that sunny autumn morning, back and forth across the hundreds of square miles of sorghum fields north of Lake Vicente Guerrero. I asked George to take us down to see what they were looking at and realized that, a dozen miles apart, both were patrolling the fields' hackberry borders. Those tall hedgerows would be full of finches and sparrows, but I guessed the falcons were more interested in the mourning doves that were sure to be there too.

I'd started on a theory that the helpless pigeons our peregrines had caught on Padre might have prompted them to look for doves, then realized that was ridiculous. It was what a lot of us do: posit some reasonable-seeming motive that later turns out to have nothing to do with why a wild creature does what it does. That was probably even true of the primal quest I kept seeing in these young peregrines' migration. Winging along below, neither falcon felt herself embarked on some desperate southward journey; each was simply exercising her newfound freedom of movement through the air, hungrily chasing smaller birds when the chance arose. The only vision of a distant Caribbean shore was mine.

Yet my larger concept reflected the truth

of these babies' situation. Oblivious as our peregrines were, their survival really did depend on being able to complete the enormous journey they had unknowingly set out upon. Unfettered by the human burden of consciousness — of the future, of the slimness of their chances — as the day wore on, flying free and feckless across the fields of northern Mexico, they moved up toward the big Sierra Madre Oriental for the night.

On our charts, the mountains' steep topography showed they were a southern extension of the Rockies, and I wondered if our youngsters might have chosen to swerve 70 miles inland, away from the Gulf, in search of the same montane wall they'd been following south since Alaska. Yet the next morning both girls headed the other way, back to the sea, which seemed more sensible because of the profuse birdlife that was vulnerable on the coastal flats. Then by afternoon, miles apart but equally indecisive, they again wandered inland toward the Sierras. Finally George got disgusted with so much equivocation and declared that the best place to keep track of both falcons' trajectories was a landing field 5 miles east of Ciudad Victoria, right in the center of their two opposing objectives.

Stuck off on a side road, Victoria's airfield was surrounded by slash-and-burn fields interspersed with tropical dry forest, and while Vose went to replace the two quarts of oil sucked up by '469's imperfect compression rings, I meandered off into the woods. A flock of white-collared seedeaters and black-headed grosbeaks filtered through the thorny brush, beneath whose tangled branches two spotted towhees toe-kicked and back-scratched in a mat of fallen leaves not yet softened by autumn rains. Like windblown scraps of butcher paper, a family of brown jays sailed in, yelling, "Hey, hey. Pow! Hey," to one another as they flapped loose-winged from tree to tree, including one bare hackberry from which a male rose-throated becard sang with optimism above forest waiting for precipitation already weeks overdue.

I was wondering if the becard would breed after the rains, but I had to drop my binoculars because down the path hurried a flustered chap in a blue leisure suit. Clutching an oily spark plug, he introduced himself as Adolfo Cisneros.

Cisneros's English wasn't bad, and he

explained that he had made an emergency landing here because one of the plugs in his Cessna 182 had given out. Since we were in a somewhat similar aircraft, he thought George and I might have a spare. He was right, since, along with his six or eight bottles of gin, Vose kept a box of parts under the backseat. Cisneros and I dragged him out of the flight office, where he was going through our falcon story in partial Spanish. With a fresh plug, we headed for the ailing 182.

It was more than just ailing; it was stricken. As George lifted the right-side cowling, in place of the engine's upper, forwardmost threaded plug hole was a ragged opening that detonation had blasted right through the roof of its combustion chamber.

Cisneros, whom we were coming to see was not a mechanic, looked expectantly at us.

"How'd this plane get here?" Vose demanded.

With hand and arm dramatizations, Cisneros showed how he had brought his crippled plane in. Then he gestured for us, por favor, to fix it.

"Mister, this thing's a goner," said George. "Cylinder head, anyway."

He reached in and, looking back over his shoulder, waggled our plug back and forth in the space. Then without a word he handed it to Cisneros. In Vose's universe that was eloquent testimony, but neither he nor I yet grasped the essence of Mexican aviation.

Half an hour later, we had the charts spread on '469's broad tail, trying to guess where our peregrines might go next, when we heard the roar of Cisneros's big Cessna. It was a strange, semi-muffled clatter, yet, leisure suit hunched up behind his shoulders, Adolfo gave us a cheery wave as he taxied by and, after a slightly downhill run, labored his plane into the air.

The kid from the flight office had come out to watch, and I waved him over.

"Rags and epoxy," he smiled. "Your spark plug is in place."

George was one notch shy of enraged.

"You ever do that before, Son?"

"Ay, no!" the boy shrugged, "But surely it will work."

By evening our peregrines had again moved back to the coast. We could tell they hadn't roosted because periodically one of their signals would move, although in no particular direction, and not far. Behind them, Vose and I passed the little fishing village of La Pesca, which, incongruously, had a paved runway.

"What you reckon they got that for?" George said, since besides the landing strip there were only weathered board shacks and a dozen rough skiffs pulled up from the water.

"Para los pescadores norteños," explained the tallest of the five young Martinez brothers, all of whom had swarmed out to the plane as soon as we landed.

"Para los pinche gringos," snickered the second-tallest Martinez, who was unimpressed with '469 since the landing strip, they told us, had been built by Yankee fly-in fishermen to receive their far-better aircraft. But falcons were something new to the boys, and after I'd showed them color 8 x 10s of peregrines, they agreed to help us look for our birds.

La Pesca had the same fish-and-algae smell as Padre, and the mudflat behind it was as empty as the Laguna Madre, so with our receiver on SCAN, Vose and I pushed down the Skyhawk's tail and rotated the plane in a complete listening arc. As we swung around the compass we picked up a distant signal from our first young falcon, then a stronger pulse from our stronger girl, who was closer, and with the day winding down I told George I thought they were unlikely to go any farther.

I still wanted to locate them, though, and since none of last spring's peregrines had slept in the open I guessed these two would look for the shelter of trees. The boys' uncle drove me over a raised shell road that crossed a salt-grass marsh, and on its mainland shore I got out and hiked.

Red-winged and Brewer's blackbirds were coming in for the night, and beneath a dense bush that broke my silhouette I sat on a gravel bank to check their flocks for a flicker of falcon wings. Yet through the binoculars there was only a turbulence of blackbirds. Layer on layer, tiers of flying birds slid past my lenses, moving one in front of the other like opaque screens. They were only blackbirds — flocks whose members might have hatched in the cattails of any roadside ditch in North America, but the intricacy of their evening maneuvers was spellbinding.

Above me, an almost solid mass of a thousand streamed across the sky, headed for a collision with an oncoming flock of a few hundred. It felt like I should duck. But without an erratic movement or sidestepping wingbeat in either group, the flocks flew right through each other. For two or three seconds their combined numbers darkened the still-bright sky. Then each troop filtered itself out of the denser shoal it had just been part of and continued on its way.

In five minutes, everything was over. As the sun fell beyond the horizon, like emptying sacks of ebony grain dark streams of blackbirds poured down into the reeds. There the cacophony dimmed, but before it died a new sound came floating over the marsh.

It was the bugle of sandhills — returned now, some of them, from nesting grounds in Asia — bringing back their Siberian-born young to winter in these Mexican lagunas.

Pealing out of the three-foot-long tracheas coiled on top of their wishbones, the cranes' hollow voices were eerily melodic, with a spooky, antediluvian feel I'd recognized decades before I learned that avian fossils like those found near Hell Creek define cranes as among the most ancient of living birds. All that mattered, though, was that here on La Pesca's tidal flat, out of the clouds this autumn's first sandhills were careening, drifting, side-slipping down.

With a concussive clamor of their heavy wings, one by one they lowered their spindle-legs into the salt grass, found firm footing, and waded out into the shallow lagoon. There, where neither wildcat nor coyote could reach them, each crane would stop, stilted above the sunset-shimmered ripples of its own passage, and raise its beak to the sky, bugling to others still aloft in the night with calls that were not only the oldest but, I felt sure, the loveliest sounds a bird could make.

28.

Policia Militar

In the morning, George and I were glad to find that, unlike Amelia, our falcons hadn't left their roosts before dawn; even without rotating the plane we saw on the receiver that they were still nearby. The Martinez clan had shown up for our early scan, and after I'd run both young peregrines' frequencies for them and described each bird in detail — our rail-thin, first-caught adolescent female, and the husky girl I'd found later — I punched in Anukiat's #.973.

No longer back on Boca Chica, Anukiat had to be nearby; as his beeps blared out of our speaker I dived out from under the Skyhawk's wing, binoculars in hand.

I didn't need them. To the north the estuary's icing of shorebirds was starting to explode. As if detonated by some silent oncoming artillery, every hundred yards a shimmering geyser of waders — willets, yellowlegs, godwits, turnstones — was throwing itself into the sky. As each rising fountain of birds came streaming down the beach, peppering our ears with their cries, another, closer flock would suddenly erupt — for in the distance, on casual wings, came Anukiat.

Just behind the scattering shorebirds, he scanned at the panicked throngs and chose not to attack. Then he changed his mind and folded himself into a shallow, downward rush. Beneath him, the plumes of rising waders shattered into atoms. But like a spaniel splintering the surface of a pond to fling the spray, he was only playing. Without unleashing his long, deadly toes, he pulled up, 80 feet above his shadow. I could see his snowy, black-mustached face and the yellow rings behind his eyes as he came down the beach. Then he noticed us, jerked his head in surprise, and as if made from helium rose without a wingbeat and slid gracefully out to sea.

All of us were speechless. Focused on the spot where his small black crescent had vanished, the Martinezes stood riveted.

"Está Anukiat," I said.

But back in the plane I saw that it was not. Anukiat's signal was still a mile back up the coast. But an unmistakable, white-cheeked *tundrius* had flown past, headed down the Mexican shore, and after the

Martinezes had helped me repivot '469 I told them that both our young females were also still close by, and the boys held out their hands for my binoculars.

Searching the horizon through the lenses, the second-oldest and most solemn Martinez announced, as though he'd seen them, "Sus niñas."

His big brother pulled away the glasses, looked for a second, and laughed: "No. Las hijas: Gorda y Delgada."

And with that, the two young females became our peregrine daughters: plump Niña Gorda and skinny Niña Delgada, and when Anukiat still had not appeared by midday, we went after them.

Ninety miles south, beyond the lower end of the Tamaulipan Sierras, Niña Gorda, who by now was far in the lead, sailed out over a brushy wetland pied with egrets. It was a new kind of terrain for her, and less than 100 feet up she was within easy striking distance of any prey she saw, though I doubted she yet knew that she would never capture a bird flitting above so much reedy cover. Still, she might have just been looking, gazing down at country she was seeing in a way no human ever will.

From her altitude, George and I could pick up nothing but a swept-by mélange of ocher and olive-brown, unfocused as a slow shutter-speed photo of tumbling rapids. But peregrines' eyes set them down right in the middle of what they see — steady, despite their speed of flight, on every leaf and blade of grass. It is a kind of sight unattainable even to humans stationed in the midst of that foliage, because falcon eyes are able to instantly magnify — instinctively flaring into focus — any distant flicker of movement that could be potential prey. And through those vastly sophisticated lenses I imagined Niña Gorda absorbing all the intricate topography of this new land: devouring, with every wing stroke, mapfuls of precisely delineated terrain.

I wondered how much of that she could recall. Then Amelia's perfectly timed descents came back — how, almost every evening, she had dropped decisively onto a high bluff above a major watercourse. Like pulling into a familiar motel, Vose had noted, and it seemed probable that these young travelers were laying down similar imagery, making mental maps that in the future would let them find their way along this route again.

Over the southwestern horizon, an up-thrust breadloaf of vine-draped limestone protruded from the lowlands, and Gorda headed directly for its barren crest. I told George the little round peak might have resembled the stony headlands of her birthplace, and 4 miles out from the summit, over the village of Magiscatzin he put us into a big listening loop. Gorda's radio swept up to the rock's crest, then halted. I tipped the receiver's window to show Vose her numbers.

He nodded. "Getting landmarks."

After only a few minutes she continued south, leaving Delgada even farther behind. But Vose was still thinking landmarks. For years, he said, he'd had a standing bet with other fliers. You could unblindfold him in a plane, anyplace in the lower forty-eight, and with just a compass, in an hour he'd know where he was.

"Took forty years to get to that point. And I still trust a decent map more'n any instrument but that little magnetic needle."

During our next stop, while George went to get the weather I was the one checking '469's controls, and once again I was struck by how spindly were the cord-like steel cables that ran back to the tail's rudder and out from our steering yokes to the wings' banking ailerons; on these simple wires, and on our hollow, thin-skinned aluminum flight surfaces, our lives depended.

Then I remembered my veterinary training. It was a course in wildlife medicine for those of us whose animal-care facilities could not afford the fees of practicing vets, and how big birds — the hawks, herons, and vultures we tried to help with emergency surgery — were put together. Their flight, too, depended on a thin external shell, a tightly joined sheet of feathers that formed a planar foil to catch the wind. Knitted into a single lamination of cross-linked keratin, those birds' big feathered sails were also controlled by strings. Long, yellowish tendons ran out along their owners' hollow humerus and ulna forearm bones to the vestigial fingers that, delicate as those of a pianist, commanded all the boundless intricacy of their flight.

Vose and I used the same simple cable-pull systems, but almost everything beyond that was mental, a component of the still-forming minds of our young peregrines; almost instinctive, for decades, in George. Neither the falcons nor Vose had attitude and bank-angle gyros, radar-direction beams, or satellite-coordinated autopilots. Those who navigated that way flew differently — automated conductors, Beryl Markham called them in *West with the Night* — and to myself I smiled, knowing that when people thought back on those who flew with the wind in their souls and the tactile power of slender wings immanent in their limbs, though they never heard his name, they'd be thinking, too, of Vose.

Down the coast, a big port city was coming up, and the Skyhawk's microphone came to life long enough to get us landing clearance. Built before Mexico's oil boom swallowed the land along the Guayalejo River, Tampico's airport was wedged between refineries and tanker depots, but on rollout down its long concrete runway, I saw a familiar flicker.

Falcon wings. More colorful than even those of a male kestrel, they were low, swooping up only to the top of one of the taxiway fence posts.

George also spotted the movement and braked to a stop, asking if that wasn't some kind of peregrine. It was a falcon, but an aplomado, named for the lead gray of its back. Fearless as a banty rooster — and about the same length — the aplomado was swallow-slim, and the mantle for which he was named was the only dull thing about him. The rest of his hot-hued coloring seemed geared to this land of overflowing hibiscus, plume-tailed quetzals, and Diego Rivera airport murals. Above his black-and-white-barred tail the aplomado's burnt-orange belly and snowy breast were separated by a theatrically broad black waistband.

No more than 30 yards beyond our whirling propeller the aplomado perched on the fence, watching us from below a gold-striped crown that gave him a severe, wasp-like aspect, and his orange-rimmed button eye stared intently — hopefully — as I motioned to George to pivot the Skyhawk.

"Waiting for bugs," I explained as our prop-wash flushed a spray of grasshoppers out of the tall grass. The little falcon darted off his perch to snatch a fat, orange-winged lubber. Back on his fence post, he bobbed his tail and, with the hopper held between his toes, squinched shut his eyes and reached in between the spiny black legs kicking at the sides of his face to snip each limb's horny stalk. I grinned at our partnership: the tiercel got a big bug, and through the glasses I was able watch him eat it in as much detail as if he'd been perched on George's knee.

Aplomados once lived throughout the grasslands of the southwestern United States. Most of them have been gone now for nearly a hundred years, but along Mexico's eastern seaboard our aplomado was one of many: at Veracruz, Villahermosa, and Chetumal these long-winged hunters patrol the weedy airport verges waiting for the mechanical wind of aircraft to blast out their insect and small-bird prey.

As we pulled in to tie-down, it was nice to see that the surrounding industrial wasteland hadn't changed Tampico's old terminal. An anachronism among Mexico's gleaming new glass-and-terrazzo passenger facilities, General Francisco Javier Mina Airport was still Old Mexico. Built with the local yellow brick, its low-rise control tower was tunneled with narrow corridors and small administrative offices, into the largest of which George and I were immediately escorted by Aeropuerto Seguridad.

There, under rows of little wood-framed windows propped open against the heat, Javier Mina's jefe sat and sweated at a rusty metal desk. He, too, was Old Mexico.

"This airplane is not a scheduled flight," Raimundo Torres said in Spanish. "Therefore it must be examined."

He waved a green-coveralled worker in the direction of the Skyhawk and, entirely unimpressed with our peace-officer uniforms, pointed Vose and me at a pair of 1940s chrome-and-pea-green art-deco lounge chairs. Through the dusty panes, we watched Torres's mechanic and a partner tentatively approach '469. They fingered our stainless antennas, traced the cables that ran in through the vent holes in our Plexiglas windscreen, and exchanged looks over the leather-cased receiver wedged between the Skyhawk's seats. Finally, after poking through Vose's suitcase, they returned to the office.

Neither mechanic had gotten out a word when, without a glance in our direction their jefe sent them back. George brushed a speck from his crisp officer's shirt and gave me a skeptical look.

Back at the plane, the director's men went through the same routine. They looked deeper into both our bags, but once again their report was short-circuited by their boss's wave. On the third trip, they just stood by the Skyhawk. Finally, one pulled open its engine cowling, stared at its greasy little motor, and because they were mechanics, checked its oil. We were down two quarts, so they topped us up.

After that they came back upstairs. This time their jefe listened briefly, but as he raised his hand to send them out again George caught my eye. The director's left-hand desk drawer was open and, face up, I could see an American twenty-dollar bill.

"We got to cover that," Vose whispered.

"Cover?"

George rolled his eyes. "Drop the same amount in there. No more."

Five minutes later, a smiling Jefe Torres shook Vose's hand as we climbed into the plane. Then he glanced back at his two helpers, standing shoulder to shoulder, and darkened.

"Diez dólares más," he growled, meaty hand resting on George's lap. "Para servicio. Dos petróleos."

At our little hotel outside the city I woke early. It was the second night I'd had almost the same dream, and what I had seen in sleep was nostalgia: the lost wildness of Mexico's mangrove-tangled eastern coast that as a boy I'd known from botany trips to this area. Nearby was the little town of Altamira, namesake of both the canary-breasted Altamira yellowthroat and the Altamira oriole, whose hanging-basket nest — sometimes suspended, back in southern Texas, from an electric wire — was an intimation of the Tropics. But the surrounding marsh had been gone since the seventies, its standing water drained into the Gulf, and in the wetlands where those black-throated orioles had foraged, cracking open insect-riddled twigs and branches with their unique bony palates, huge gas-powered generating and refining plants were springing up along miles of newly bulldozed red-dirt roads.

Nevertheless, like Amelia, Niña Gorda had taken us over a frontier. The day before, Vose and I had crossed the Tropic of Cancer. That seemed crucial. The Tropics — home of rain forest, of thick jungle rivers opening onto bird-rich coastal flats — was where I'd imagined our falcons' winter home to be.

George was up and hungry for the manteca-fried tortillas that went with every plate of scrambled eggs, but to continue our telemetry we were going to have to change strategy. In less than a week, I pointed out, we'd spent nearly a quarter of our cash, most of it on the gas it took to log up to 500 miles a day in continually

481

trying to locate Anukiat and Delgada — both of whom, as far as we knew, were still somewhere back up the coast.

We were going to have to do more listening from the ground, Vose agreed, and in order to do so he had gotten a kid at the Tampico flight desk to pencil in a bunch of unmarked airfields, off to the south, where we could land and listen to Gorda's out-front transmitter. With that in mind, an hour later Vose and I found her winding through wooded coastal foothills, and in hopes of getting a glimpse of her long wings, we floated down over the tops of dense jungle trees. It was our first demi–rain forest, and I was glued to the Skyhawk's window when I saw big, dramatic birds in flight.

"Parrots!"

George leaned us over, and there was my fantasy. Emerald backs shining in the early light, two pairs of large parrots flew side by side, skimming the berylline canopy with the quick wingbeats their heavy bodies required. From their stubby tails I knew they were Amazons, probably red-crowned, though as they skirted the gray-branched crests of the tallest acacias I couldn't make out their scarlet caps and yellow-tipped tails. But to me they were a symbol.

Eight weeks ago, our Niñas and Anukiat had perched on icy arctic bluffs; now Gorda was flying over tropical woodland in the company of shamrock-green parrots. As George and I swept across a carpet of intertwined forest crowns, I told him how much I wanted to see those big trees up close, from the ground — see their lianas and strangler figs, their clinging, red-flowered bromeliads and climbing monstera.

For that, one of the marked-in landing fields on our map looked perfect. Deep in Distrito Tuxpan, Cerro Azul — Blue Hill — lay in the center of an upland where we would have line-of-sight reception toward both the Gulf and the Sierra Madre to the west. Checking the chart, George said it should be easy to find.

This time, though, Vose's faith in maps proved unwarranted. Incised among the trees, Cerro Azul's runway shot by on our first pass, giving us only a glimpse. George noted that it was long enough, however.

"Bunch of buildings, too. Maybe we can get a snack."

He came around on a downwind leg and cut the throttle. But as we touched down there were soldiers running toward the plane. More were pouring from every door of the olive-drab barracks that lined the airstrip, and Vose and I exchanged horrified looks.

"Military base," he said, and in the next second the Skyhawk was surrounded. Angry, Indian-brown faces under olive caps pressed against our windows. With our antennas, I gasped, they think we're spies. Then my door was yanked open by a commando with an automatic rifle.

Behind him, poised in an enclave of his troops, stood a young soldier, Aztec/Mayan like the rest, but wearing a big copper medallion.

"Soy El Capitán Mendoza," he shouted. "Commandante Temporal del Base Militario de Cerro Azul. Porqué están aquí?" (Why are you here?)

I fumbled for words. "Científicos." Scientists. "Sigiuendos un halcón, con radio. Desde los Estados Unidos." (We are following a falcon, by radio. From the United States.)

It was hopeless. Vose and I were hustled over to the open side of a partly built masonry building, where we stood, more or less the focal point of twenty ready, but not raised, carbines.

There, nobody moved. Least of all us. Beyond its runway the compound was only a cleft in the forest, its perimeter secured by a mossy stone wall that impounded the measureless humidity of the Tropics.

I was shaking; then I heard George whisper, "Easy, Mister. Somebody's coming."

A rusty American compact bumped down the narrow dirt track, and from its rear seat a rumpled officer stepped out.

"A general," muttered Vose. "One star."

Short and heavyset, the general strode past '469, whose scanner, as if in testimony to our nefarious intent, still beeped with Gorda's waffling signal. In Spanish, the general fired a barrage of unintelligible questions. George and I shook our heads. Then he came closer, glowered at each of us, and leaving the plane as untouched as a murder site, stamped back to his vehicle.

Other than that, everything was still. Over the barracks a cookstove haze hung halfway to the tops of the jungle trees, and in the distance, diving from forest edge to forest edge, a flock of long-tailed conures cut across a corner of the clearing. Thunderheads built in the stagnant air, and surrounded by small, scowling faces, I wondered if we were all going to stand, impassive as Aztec carvings, during the coming downpour.

But before the storm another car rolled up. It was a better car, a polished four-door Nissan, and its driver was the same single-star general who had just been chauffeured away. This time, he was driving with passengers, and as we waited for the Nissan's rear doors to open, as if remembering his duty the general jumped out, scrambled around, and, standing at attention, swung open the right-side opening.

Only then did the real Comandante emerge. He was tall, mustached, and confident; a mop of curly black hair swirled from beneath the band of his visored field cap.

"Two stars," George hissed.

Trying to explain in Spanish why we had landed an electronically outfitted surveillance plane on Cerro Azul's restricted runway had already proven to be out of the question. So, trembling as little as possible, I held my aluminum briefcase in both hands, stood at attention, and waited for the commander to come to us.

When he did, there was a long minute of sizing up. Epauletted military khakis chest to chest with my chocolate-and-white Highway Patrol polyester. Black comandante Ray-Bans eye to eye with our silver-mirrored Texas peace-officer shades. Then, from the anodized case containing our bogus credentials, I withdrew the Mexican Presidential Envelope. Layers of fake gold leaf. A gilt perimeter of Texas House of Representatives Statehouse seals, two big bouquets of diplomatic ribbons, and a notary public imprint in red wax.

Our diplomatic piñata. From the typed card Burrell had given me, I read:

Somos envitados del su excelencia
José López Portillo y Pacheco
El Presidente de Mexico.
Identifiquese, por favor.

(We are guests of his excellency,
José López Portillo y Pacheco
President of Mexico.
Please identify yourself.)

The commander stared from me to the envelope, scrutinized its authentic presidential insignia, then slowly turned back to us. Vose, at least, with his authentic military bearing and still-starched trooper's uniform, looked genuinely impressive.

All the general's officers were watching, and the next instant the commandante whirled to face them.

"Porqué no me avisarón?" he snapped.

"Rapidamente?" (Why was I not notified at once?)

Mendoza pointed toward the single-star general, but his commander cut him off.

"Nunca!" he barked at his attaché. "Nunca! Ni me avisarón o visto estos papeles!" (I never saw these papers! Was never notified!)

Then he strode away.

Keeping a wary eye on their commander's back, everyone remained perfectly still until we heard the Nissan's rear door slam, then watched as its single-star chauffeur steered it carefully back up the curving path between white-trunked flamboyant trees. After five minutes, George and I were hauled back to the Cessna, still sitting on the runway with both doors open, and moments later we were off the ground and in the clear.

29.

Corridor of Hawks

After Cerro Azul, I thought I might lose Vose, but without dispute we flew on, picking up Gorda in the Papanteca highlands. We were starting to look for another place to set the plane down when George pointed ahead.

There were chalky tiers rising above the surrounding forest. A ruin. Some sort of pre-Columbian temple. Neither Vose nor I had expected anything like this, and he nearly capsized the plane bending us over for a look. In the undersea pallor beneath still-growing thunderheads, the building's limestone blocks glowed with a pale-green patina and, left wingtip pinned to the spires' summits, we pivoted down over what I had found on the charts was El Tajín. Capital of the Totonaca culture, Mayan Sacred City of the Dead and Thunder in Storm. Around the temples there had been a city, built of wood and thus long decayed, but a city open only to rulers, politicians, and priests.

Its inhabitants had been powerful figures, with their limestone towers reaching above the highest trees, and below us now — just barely below us — was the highest of those temples. It was the Pyramid of the Niches, whose 365 recesses, each representing a day of the year, were set into steep stone faces that looked so much like the falcons' ancestral cliffs that I wondered if our arctic Niña would try to land. But with a storm blowing in from the north, Gorda stayed in the air ahead of the weather, oblivious to the crumbling grandeur of the monuments.

Then the clouds closed all the way in and gusting winds shook '469 so violently that George turned back to the good-size town of Poza Rica, where rain kept up most of the night. By dawn the wind had died, and as we rose into a sky scoured of color by the storm, I could see how narrow our ribbon of coastal plain had become, compressed by the mountains pressing in from the west. Eighty miles to my right, in purest clarity stood Orizaba — Mountain of the Stars in native Nahuatl. Spotless in perpetual snow, it loomed above the continent's volcanic spine; off Vose's wing lay the midnight-blue Gulf, flecked to the horizon with whitecaps.

George had been fiddling with the radio, which had died again during the storm over Poza Rica, but muffed in his earphones he followed my right–left gaze, taking in both flanks of the spectacle in which we hung suspended, and nodded in agreement. I snugged on my own headset, got Niña Gorda on the scanner, and thought how, even long before falconry, men must have been drawn to birds of prey because of the lavish world in which they lived.

But few of those visionaries could have imagined what now lay before us. Ahead and to the left, in the direction of Gorda's signal, the camel-humped peaks of olive-green Punta Mancha thrust out into the Gulf, and as we drew nearer, a haze began to coalesce in the passage between their mossy slopes. Moments later Vose and I drilled into that darkening mass, and suddenly on all sides there were swirling, flickering flecks of life. They were birds.

Throttled back just enough to keep our altitude, '469 was surrounded by billows of avian life. Mostly no more than just-glimpsed specks of pigment, almost all the birds were too small to identify — part of the millions of northern migrants making their way down the continent's central migration corridor — though we could see thousands of larger soarers, too, almost all of which were birds of prey.

It was a raptor flyway! I yelled at George. Like Hawk Mountain. Vose shook his head at the unexpected obstacles and began to climb, but the loft coming up under our wings told us there was rising air for our companions, too, and over Mancha's rocky beach we found ourselves rising alongside a revolving kettle of gliding hawks. Some were Swainson's, but most were the smaller, band-tailed broad-wings.

"Lot of whatever-they-are to get in the way," George observed. "Hazard to aviation."

But they weren't. Though neither Swainson's nor broad-winged hawks are agile flyers, those that found themselves in front of our propeller gave a single startled flap, then, as '469 swept up behind them, peeled off into a side-slipping fall that dropped each obstructing individual cleanly out of our path. I thought we would pass the flock in seconds, but for miles Vose and I flew through a thinning, then thickening vapor of big brown birds, every one of which gracefully folded its wings at our approach and moved deferentially aside, ushering us through an endless corridor of hawks.

Then — I couldn't believe it — I saw a peregrine. A swift black arc, it scythed up through the drifting buteos, bursting out of their upper layer like a dolphin exploding through the surface of the sea. In an instant it had ripped away into the sky, but the next falcon I saw — big, chocolate-backed, and like Amelia part of a rising kettle of vultures — flared off from '469, then came back. Fearless in the air, she dug her wings into the tempest of our wake, and during the second or two she was able to hold pace I saw that below her brown head her throat and breast were mottled with umber.

As the falcon cut away, George tapped the receiver, but I shook my head. She was neither a *tundrius* nor Gorda, though our foremost Niña was nearby, her signal coming strongly through the receiver. I scanned for Delgada and Anukiat, got nothing, then punched in Amelia's #.759 and showed it to Vose, who put one hand on his headset and listened for a long, fruitless while.

Yet all around us the soaring hawks were phenomenon enough. Having made their way down from the forests of the eastern

United States and southern Canada, the constellations of perpetually gyring broad-wings had arrived on the Texas coast about the same time as our peregrines, though their journeys had little in common. Hunters of mice, frogs, and insects during summer, broad-wings on migration become economy specialists. They are unable to capture flying birds and refuel themselves on the wing like peregrines, so they seldom feed during the weeks they spend moving south. It takes that long to get to their winter homes, for although some of the individuals passing by our windows were going only as far as Central America, other broad-wings would soar on to the *llano* plains of Venezuela and Colombia.

None had sufficient fat reserves to let them flap the whole way. Their only choice, therefore, was to drift slowly southward, carried by the warm winds of autumn, away from the winter that was soon to grip their northern breeding grounds. The shortest route to Central or South America was straight across the Gulf of Mexico, the path tiny songbirds follow every spring and fall. But water doesn't generate the rising columns of air that are generated by sun-warmed land, which left the soaring hawks only this narrow channel, between the sierras and the sea, where thermal currents could take them up to the altitudes where prevailing southerlies occur.

George asked why the broad-wings didn't hunt along the way, and I told him they had no time. After roosting together, sometimes on the open ground of agricultural fields, at dawn their great communities go aloft, searching for the earth's first rising air to give their lightly loaded wings the tiny boost that will eventually lift them higher than the reach of human sight. There, like the bands now passing above and below our wings, traveling broad-wings spend every sun-warmed hour in slowly rotating galaxies, feeding on nothing more than an occasional insect or rodent picked up at dawn after a night in a field.

Theirs was a slow voyage, epic in its way, but compared to the Swainson's, the broad-wings were almost home. Swerving away from slowed-down '469, the Swainson's were western raptors, birds that had arrived from nesting grounds spread across the whole sweep of the Great Plains. Yet here in the horn of Mexico they were still no more than a third of their way along a largely unknown route that would take them across the entire Amazon basin to the grasslands of Uruguay, Paraguay, and Argentina.

It was a hazardous trip, more so because even on those distant plains the Swainson's found no haven. No longer pampas, the Argentine countryside looks like Kansas: endless rows of corn and sorghum, empty of natural raptor roosts. There, feeding on the organochlorine-permeated grasshoppers and mice that manage to survive the grain farmers' periodic poisonings, thousands of Swainson's hawks succumb to chemical toxins, while without other trees in which to roost, many of the Swainson's sleep in the windbreak eucalyptus and casuarina arbors that shelter every plantation's headquarters — where, according to global birder Scott Weidensaul, they are regularly gunned from the branches by bored ranchhands.

That night Vose and I hired a cab to take us all the way downtown, where we got a room on Veracruz's waterfront. During dinner, I told him how flying with the fasting hawks had renewed my worries about Delgada, somewhere far behind us. Unlike the soaring buteos, falcons have to work for every mile, and our emaciated first Niña carried nothing to sustain her.

George said he wondered how long he could sustain himself on the Mexican chow we'd been getting, because his innards hadn't been the same since we left Brownsville, but that he could maybe see how the concentration of hawks we'd just come through didn't eat much, even after flying all day; what he couldn't understand was how they went so far without stopping for a drink.

I replied that as far as anybody knows, broad-wings and Swainson's make the whole trip on just their bodies' retained moisture. Peregrines might be able to do that too, but if they have the choice they usually go down to water first thing in the morning. I'd seen California fledglings, wrinkled toes submerged in riverbed shallows, dip their beaks, then tilt back their heads like drinking chickens. Sometimes they even walked in deeper to bathe, splashing and flailing their wings, and I was sure that being able to go for an early dip was part of the reason Amelia had always stopped to roost near water.

If they were feeding, I said, peregrines might not need to drink at all. Their physiology didn't demand it — not like a mammal's — because we get rid of metabolic waste by excreting urea, a semi-solid that takes a lot of fluid to flush out as urine. Birds don't have urine. Or a urinary system. Every bit of their metabolic by-products goes out through the single big cloacal tube they have in common with lizards and snakes, alligators and turtles — as well as with their dinosaur ancestors — through which they excrete not urea but uric acid. Without being further diluted uric acid can leave the body in either milky droppings or the little tubular scats of waterfowl and galliform birds, and most of the minimal water this process calls for can come from a bird's food.

Among falcons, that may mean the blood of prey.

George said that couldn't amount to much. Not from tiny little songbirds. He was right, but when their prey is larger its blood can be substantial, like that of a pintail drake I had seen a peregrine ride to the ground; after severing the duck's neck, for almost a minute the falcon had lapped every drop of its arterial pulse.

Vose put down his fork and looked at me funny.

"Mister, I guess I'm glad you know all this. Just don't go telling it to anybody you want to help save these things." He thought a minute. " 'Specially at dinner."

Before daybreak, through our open windows I heard the sweet, determined peeps of migrating barn swallows. Barely interrupted by darkness, yesterday's southward march had begun again. In front of the hotel, Veracruz's miles-long bay was lined with gnarled Australian pines, and as the taxi took us along the bordering Avenida Tintorera I noticed that some of the trees held merlins. Their slim, big-headed shapes were almost obscured by the silvery needles, and I'd never have spotted them except that every so often one would arrow out, swerving among the park benches like a big predatory insect, intent on the quay's scattering packs of house sparrows.

It was a raptor concentration I'd never have imagined, for in North America merlins are found either by themselves or in pairs like the couple I'd seen perched amid Denali's black spruce. I pointed out the little falcons to George, wondering if I could again be seeing either member of my Alaska pair. He said their erratic chases reminded him of the jumpy-flying sphinx moths that came up to the house lights at his ranch, but now that I'd gotten him involved with three more peregrines he didn't have time to be interested in other falcons.

We needed provisions, so at the corner of Tintorera and the Southwest Highway we stopped at a supermercado, open early, beyond whose big glass doors I was hit with a wall of sound. Disco was alive in Mexico, at dawn, in the shopping aisles. Between bouts of teen patter and product endorsement spieled by a coiffed CD spinner set up in front of the cash registers, vintage Madonna blared across the shelves. But the shoppers, mostly young moms with babies on one or both hips, were oblivious to both the music and the DJ's advertising come-ons as they trudged up and down the dusty corridors.

George couldn't find anything he wanted to eat, but a gang of Chiclet kids attracted his attention, and reaching for pocket change, with a broad smile he started toward them. They scattered.

"Well, what in hell?" he said.

But the disc guy was hip: pointing at our FALCON TRACKING TEAM shoulder patches, he laughed.

"Los halcónes: Policía de mucho fuerza."

Out on Mexico 140, the airport had more runway than we needed, and as Vose climbed toward Isla Arrecife Cabeza I turned on the receiver. No signal from Gorda. George wagged our nose right and left to give the antennas more reception, but all I had was static, so he eased us into a wide listening arc.

There was nothing, but as our circle grew, on its south side we came over wetland savannah. Downstream from the Oaxacan border, the big, slow-moving Rios Camaron, Papaloapan, and San Juan came almost to a standstill, flooding 200 square miles of marshy pasture. Before its original forest was cut for cattle, this was one of the most diverse woodlands in the Neotropics, home to nearly 700 bird species — more than occur in all of Guatemala or Honduras. Opened for grazing, the old meanders of the Lagunas Popuyeca and Pajarillos lay incised between bristling banks of red marsh grass where humped Zebu cattle, adapted to heat, biting insects, and always-soggy hooves, clumped together like pale-gray cobbles piled at the foot of the occasional big kapok.

Because the columnar trunks of big kapoks rise so far into the sky, the huge plants, also called ceiba, were "trees of heaven" to the Maya. Mature mahogany and the biggest matapalo figs reach almost as high, but after seeing my first giant kapok growing in thick jungle I could understand something of what its kind must have meant to native Central Americans. Like some supreme architectural pillar, the tree's fluted spire of living wood had soared up, branchless, through tier after tier of intervening forest canopy, striving for the shadeless heavens.

There, above the uppermost story of the surrounding trees the largest kapoks spread their flat umbrella brims so high that only from another kapok's crown — or from the uppermost platform of a Mayan temple — could their heavenly summits be seen by men, who, before they built their lofty churches, lived like termites along the shadowy trails 100 feet below the sun.

Now, standing alone on a sea of newly grown grass, the Camaron's big trees were all that was left of the towering woodland in which they sprouted. Uncut by ranchers for the patch of shade their broad summits offered livestock, both kapoks and matapalos were characteristic rain-forest remains. Worldwide, peculiarly tall, willowy trees stand at the edge of agricultural fields and housing developments built on the borders of jungle woodland. Capped, far above the ground, by smallish crowns,

the trees look gawky and out of place because only in the anachronism of their
shape is the immensely tall, thin-stemmed
vegetative world in which they grew still
present.

In the wet Tropics the sun's energy is so
completely devoured by the forest's upper
layers that almost none of it reaches the
woodland floor; as seedlings rising from
that shadowy litter, forest trees can survive
only by fighting, every second, for the
light. It is a long struggle, because in a sort
of suspended adolescence rain-forest saplings often wait more than a decade for the
sunny opening left when one of the shade-
spreading patriarchs standing above them
crumbles — most often from the thousands of pounds of vines and strangler figs
that, also fighting for the sun, have
swarmed up its trunk and spread along
every high branch.

Although rain-forest profusion was the
recent past of this cut-over marsh, below
'469 thousands of slow-flying cattle egrets
now floated above miles of inundated
grassland. It was country open enough for
a peregrine to hunt, and by mid-morning
George and I located Gorda's transmitter,
motionless near a copse of thorn brush.

That meant she was either dead or had made a kill, so Vose spiraled us up, climbing as slowly as a kettle of broad-wings. By noon Gorda was in the air again, and, 10 miles behind, happy to have her back on the wing, Vose and I also headed south.

But our long vigil had consumed nearly half of one wing-tank's fuel. We had no choice but to spend more time on the ground, and beyond the marsh's southernmost *rio* George began letting down over the agricultural fields around Lerdo de Tejada, where the moist black soil grew sugarcane.

Like broomstick-stalked wheat, for miles a 7-foot-tall mat of golden-green cane occupied every acre. Many of the fields went on farther than we could see, and I was wondering how their owners got the stalks to market when a dirt airstrip appeared below my window. Cut right through a stand of cane, the rudimentary landing field was narrow, but level enough to set down on. Wary of more trouble, George circled while we scrutinized the surrounding terrain. The only sign of potentially offendable humanity was a brace of dirt ruts that, a mile away, climbed a low rise, so Vose bumped us onto the ground, where, with Gorda still clear on the scanner, the rough surface dragged us to a quick stop.

With the prop shut down, the electric cheeps of Gorda's transmitter were the only sound because, along with every breath of air, the narrow strip's walls of cane cut off every trace of noise. In the heat, George and I had pulled out our camp stools and were just settling in the shade beneath '469's wing when we both jerked to our feet.

A sharp, angry voice had barked from somewhere behind us, and, turning, I realized that what I was staring at was the shiny black barrel of a .45 automatic, held by a stump-shaped man whose words George and I could not understand. He waved the gaping muzzle over toward Vose, then back at me. Whatever the man said wasn't in Spanish — K'iche', maybe, or Yucatec Maya, but it didn't matter, because he was in such a frenzy I was sure he was about to kill us.

"Oughta move apart," Vose hissed through locked teeth, stepping away from me. "He gets one —"

The peasant's staccato command cut him off. Clutching his gun in both hands, the man stumbled over the cane furrows to his right, lining Vose and me up again. It was only then that I noticed he was wearing no pants. Behind him, a narrow dugout had been slotted into the thick stalks bordering the runway, and in its bottom were a cookpot and strewn bedding. The guy must have been asleep, which could only mean Vose and I had come to a dope field — one of the temporary airstrips scattered across Mexico's back country and used only once, we'd been told, always at night.

This gunman was its guard, and under the broiling sun all we could do was stand, trembling, as he decided what to do. His initial frenzy seemed to have calmed, but still holding the heavy gun at waist level, he edged toward the Skyhawk. Both its doors stood open, and he looked inside — trying to tell, I imagined, if the plane seemed like something he could operate. I was thankful for the complexity of the Cessna's instruments and the fact that its windshield was too high for him to see out of.

But turning his back on us, he started to reach in.

"Whoa now, Buster, you can't . . . ," George began, but the man whirled, pulled back his gun's hammer, and bared his teeth in determination.

Droplets of sweat slid down my chest,

but all I felt was chills. My friend Jan Reid, gunshot by a teenage mugger on the streets of Mexico City, told me that at the British-American Hospital a doctor trying to deal with the agony of the .32-caliber slug lodged in his spine had asked him to rank his pain on a scale of 1 to 10, with 10 being the worst he'd ever felt. Jan said it was 10,000.

Here, days from medical care, I realized the best outcome of getting shot would be to be killed outright. But, little by little, Vose and I could see this fellow's hostile panic begin to diffuse into confusion. I hoped it was occurring to him that since he couldn't move the plane, our orange and white Cessna would surely attract attention, and as if in confirmation we heard a motor. It belonged to a muffler-less sugarcane truck that struggled up the little hill in the distance. No one was coming to our rescue, but the cane cutters standing in its bed had surely seen the Skyhawk, and I hoped that might give our captor a reason to let us leave. But he seemed only to stew more intensely.

"Don't say a word," hissed Vose. "Longer he thinks, the better. . . ."

In silence, the peasant made up his mind.

"Back in the plane," he motioned, stepping forward, and George and I climbed in. We had barely cranked up the Skyhawk when I put a hand on Vose's arm. The man's face, and his gun, were just outside my window.

"Vose," I said, "he's going to shoot on takeoff."

"Get rid of us in a crash," George growled. "But . . . what else we gonna do?"

With his cocked .45 pointed in my window, the man trotted along next to '469 as Vose idled downwind. Then George locked his left brake, pushed the throttle home, and in a spray of dirt gunned the Skyhawk around, pointing down the runway into a slight takeoff breeze.

But surging over the turned-up clods meant we could barely accelerate, and our captor had already sprinted a quarter of the way back down the strip. We'd have to roll right by him to pull into the air. It was too late for another strategy, though, and beyond my right-side window I saw the man swing his pistol up and steady it in both hands, dead on our bumpy course.

George saw it, too, and we braced for the rip of a big slug.

But nothing came, and as our wheels

cleared the ground, hunched forward over the yoke to make himself as small a target as possible, Vose couldn't see the man sight down that massive black barrel, sweep his pistol along our rising trajectory, then slowly lower his gun. As we climbed, I finally breathed, pounded George's shoulders, and yelled that it was OK — we were alive.

But only after a long, sober minute did Vose let himself ease back, turn to me, and say — as solemnly as if it were an actual possibility — "Alan: never, ever, let me go back to Lerdo."

30.

Petróleos Mexicanos

Ahead was the Sierra de Tuxtla, a small coastal range shrouded in dark-blue clouds. George skirted their bad weather, and as we came around the mountains' western flank we found we had again overshot Gorda. As usual, there was no signal from Niña Delgada or Anukiat.

Not far past the mountains was a small airport at Minatitlán, so with all our falcons now behind us, we went in. The field had gas and a small cantina, where we sat for two hours during the afternoon rain — now an everyday occurrence — until we realized from my periodic transmitter scans out on the runway that none of our peregrines was going to go by. Then George and I flew back the way we'd come, and before sunset located Gorda not far behind the Tuxtla's ridgeline.

The next day she didn't move much, but after listening to her from the airport's parking apron I was sure she was hunting the grassy fields below the peaks. I was worried, though, because even when we flew for an hour back up the coast neither Delgada nor Anukiat came over the radio, and by afternoon I'd begun to wonder if this bird-rich hook in the southern Mexican coast could be the tundra falcons' final stop, the wintering ground where they gave up their daily flights to the south and spread out into individual hunting territories.

Ornithological orthodoxy said it was not. Nearby lay the Isthmus of Tehuantepec, a sedimentary infill that, geologically speaking, has only recently linked North and South America. Because Tehuantepec is all lowland, raptor researchers have long imagined it to be the gateway where peregrines from the Central Flyway swing over to join the Alaskan and Western Canadian falcons that are thought to travel down the western continental shore to South America's bird-rich littoral plain.

Yet at 9,000 feet George and I turned up nothing. As we moved west, scanning the isthmus, the lowland haze began to fade, and out past Chiapas's coastal sierras the horizon gradually opened into a clear indigo. Beyond lay the Pacific.

I motioned "back," and George angled us into a long return loop that took us down over the border of Guatemala. We had already seen the old jet fighters the Guatemalan Air Force used to patrol their northern perimeter, but from our height, in the distance I could just make out mile-deep Lake Atitlán. The region's deepest, most biologically unique body of fresh water, it was Central America's Baikal.

There, for twenty years, naturalist Anne LaBastille had striven to save the giant flightless grebes that nested in the reservoir's bordering reeds. But as she chronicles in *Mama Poc*, a 1984 earthquake cracked the loch's volcanic bed, lowering the water level and destroying much of the shoreline marsh. At the same time, dozens of local villages began pouring large amounts of pesticide, fertilizer, and raw sewage down hillside pipelines that empty into the lake, and, probably too late, both the World Wildlife Fund and the Smithsonian Institution offered financial help to save the lake's receding shores.

George tightened our bank angle, and without arousing the Guatemalan jets we again headed back toward the Caribbean where Niña Gorda, faint now even with her 85-mile-range 040 transmitter, had moved down the coast toward Yucatán. Following that heading, almost immediately Vose and I were back over tropical forest, its greenery incised by thick brown rivers flowing into the Bay of Campeche; somewhere far to the north, we still believed, were Delgada and Anukiat, both coming our way.

In an hour, I wasn't sure I wanted them to be.

We had quickly overflown Gorda, mostly because she had slowed and begun circling — behavior that might mean she was investigating a hunting site. Then I saw what had drawn her attention.

Along Campeche's shore lay rows of rectangular, perimeter-diked impoundments. From the air they looked like the flooded rice fields over which peregrines sometimes hunted on the Texas coast. But these lakes were black. Shiny, petroleum-residue black.

I motioned for Vose to bring us down.

Stretched like glistening tiles, the tanks went on for miles, and as we dropped lower I spotted each pond's white inflow pipe, capped by Tinkertoy valving that tied it to long communal supply lines. Nearby stood a metal warehouse whose lettered roof told us this was a property belonging to Petróleos Mexicanos: Mexico's petroleum monopoly, whose Pemex stations occupied every major highway intersection and from whom, since Brownsville, we had been buying aviation fuel.

"Dump sites," I said. "Where they put the sludge they can't refine."

As we swept low over the tanks, '469's shadow flushed a pale confetti of shorebirds, and to avoid them George pulled up. That gave us a long-range view of the oil wells standing on raised platforms out in the bay, from which the tanks' residue was pumped.

"Forty, forty-five years ago you could see the same kind of ponds all over Texas," said Vose.

North of the Rio Grande, petrochemical companies can no longer dispose of drilling waste in that way, but lying right in the path of North America's Central Flyway, Pemex's chain of sludge tanks could be a factor in the deaths of many first-year peregrines. Drawn to every shallow basin along their way, the shorebirds feeding below were certain to be picking up petrochemical-laced invertebrates, and among the sandpipers and dowitchers, some would sicken from the toxins, and within the acres of scoured dirt surrounding every reservoir, any sluggishly moving prey bird would be a magnet for a falcon.

Especially a ravenous first-year peregrine who was having trouble catching anything healthy and agile. I wondered if even a brief hunting foray here could transmit a significant dose — maybe an ultimately fatal dose — of aromatic hydrocarbons to a feeding hawk. Years before, I'd worked for a bird-of-prey project on the California coast where adult peregrines had been found dead from dioxin poisoning. Chemical analysis had turned up contaminated gulls, probably tainted by fish poisoned by a local paper plant's inadvertent dioxin release, and though only a handful of such gulls were found, each one contained so much dioxin that feeding on two or three would have killed any peregrine.

At our next gas stop, in Villahermosa, I told Vose I had come to a decision. The brief scan we'd had from the air was only a clue to what might be happening around the overspill reservoirs, and wherever our wandering falcons went from here, at some point I was coming back.

I wouldn't expect to find an immediate answer, but with enough field samples, Don Morizot might.

Villahermosa was a rough, oil-boom town, and George and I ended up in a low-end hotel whose maroon and olive ball-room, powered by thunderous TV music, was the only place to eat. We were its sole patrons. At a lone table set right on the dance floor, watched by several elaborately dressed ladies with scary makeup, we ate our fried steaks and potatoes. George could hardly get his down.

His stomach had been acting up, but I knew that might not have been all, so I asked him if, beyond the food, he was losing interest in our mission. He thought awhile and said that, anymore, it didn't have that much to do with me.

Next morning, at V. Hermosa's flight office Vose seemed less distant. Rolling his eyes, he handed me the latest — useless — meteorological forecast.

"I already know *today's* weather," he told the girl who gave us the printout.

She smiled, waving up at the heavy clouds it described.

"Si. Está correcto."

Actually, we did learn something from her report: back up the coast a low-pressure trough had formed over the Gulf. If it grew into a tropical storm, Vose warned, even its outer rim could demolish '469. So we ought to move out of range, even without Gorda.

A month ago I might not have been willing to abandon her — still hunting, as far as we knew, back near Pemex's sludge ponds — but she had been moving so steadily southward along the Mexican shoreline that my fear of losing her was lessening, and since there was nothing we could do to help her in the storm, Vose and I flew on, across the multi-channeled delta of the Rio Usumacinta.

Close to a mile wide and the color of rust from its rain-driven sediment, the Usumacinta was Mexico's Mississippi, and in the immense wetlands that border both its banks a hundred creeks curved serpentine swirls of absinthe and brown. Then, just inland from the Laguna Atasta a gleaming pearl necklace seemed to be set into the avocado surface of one of the bay's tributaries.

Each holding perfect placement, nine white pelicans waited for the tide to sweep schooling fish out of the laguna's largest channel; circling lower, we could see each bird paddle to hold its position in the current, lean forward to stab its bill into the oncoming stream, then withdraw its arched neck and tilt back its head to swallow, just in time for another pelican to reach in. Finally, like an Esther Williams water ballet, all nine birds flared their wings and stretched forward in unison, churning a bright foam of silent splashes.

Beyond was Ciudad del Carmen. To escape the storm, most of the Mexican trawler fleet had moved behind Carmen's barrier islands into the Laguna de Términos, and the whole town smelled like shrimp. George and I got a cab that delivered us to a small hotel whose courtyard was surrounded by a high, persimmon-colored wall fortified with broken bottles set in concrete between a double row of iron stakes.

That much armor bespoke bad times in the neighborhood, and again it occurred to me that Vose, now having left even our nearest Niña 80 miles behind, might be ready to go home. Tied down within feet of the tidal surge that lapped at Carmen's seaside airstrip, '469 was anything but secure, so as we hauled out chairs to watch the rain pour off the orange tiles of our overhanging roof, once more I brought up our keeping on.

For a long time, he did not reply, and as the febrile smell of the wet Tropics filled our bulwarked patio, down went slugs of gin and jugo fresca on both our parts. Finally, George took a breath, then just

518

shrugged. It wasn't so much about my interest in peregrines, he said, but so far he'd never faltered on any mission he felt was still possible. I knew Vose wouldn't admit our falcons' lives had come to mean anything to him, but he finally conceded that by now he'd gotten sort of used to the notion of keeping '469 locked to any peregrine we had a radio on. All we needed, he said, was a decent weather report to let us know when we could get back up the coast to find our Neenas.

By the next afternoon, as '469 waited at the fuel pumps behind four Pemex offshore helicopters, we learned from their long-distance radios that the tropical storm had gone ashore more than 100 miles northwest; only the southern edge of its rotating winds would have reached Gorda. I hoped they had blown her on toward us, away from the petroleum ponds; what the gale had done to Delgada and Anukiat — who could have been right in its path — we had no way of knowing.

But to try and find out, under clouds that hung like tattered ceiling paper, Vose climbed us into a long loop that, from 8,000 feet, would let the scanner pick up any of our transmitters along the entire coast of Campeche. We heard nothing, though, and as George continued our circle out over the sea, the fear that we might have permanently lost Gorda twined around my chest.

All I could do was stare down at the battleship-colored waves. On that gray day they looked like the North Atlantic, although below their whitecaps lay both the intricacy of tropical corals and the ancient evidence of a pivotal event in evolutionary history. On the northeast side of our listening loop was Chicxulub Crater, the submerged impact site of an asteroid.

From that 100-mile-wide undersea pockmark, 65 million years ago billions of tons of debris had exploded into the atmosphere, generating a worldwide pall of dust that obscured the sun for years. The collision came on the heels of many smaller asteroid impacts and later volcanic upheavals, and the resulting worldwide dust cloud shriveled plant life and helped exterminate, along with every other large land animal, the last of the declining population of dinosaurs. That left an ecological void — one that allowed other creatures to proliferate — among them both the already long-established lineages of early birds and the peripheral little life-forms that were mammals.

Without the asteroid, those inconspic-uous, milk-producing inhabitants of wood-land leaf litter would likely never have thrived, diversified, and ultimately be-gotten the long line of descendants that led to George and me, now circling fruitlessly in search of our Niñas over the bomb crater where both the diversity of modern birds of prey and the future of every human being yet to come were delivered into possibility.

The next morning it was only drizzling, and while I loaded '469 Vose flipped on the scanner. There, not far inland from where Pemex's piles of drilling ordnance waited to be shipped offshore, lovely, plump, strong Niña Gorda was on the wing. Vose and I almost leapt into the air, following a signal so clear we could hear it even without headsets, swinging three mile-wide figure Ss to keep our ground speed down to her 45 m.p.h.

Below lay the dry forest of Yucatán, and for the first hour I was entranced. Then George reached down, twisted a silver disc on the floor that I'd never paid much at-tention to, and after a minute settled back with a relieved sigh.

"What was that all about?"

"Fuel selector valve," he explained. "Doesn't always work, you know, switching from right- to left-wing tanks. Now we're all right: over on the full side."

He'd never worried about that before, I said, but Vose pointed out that, apart from steep peaks and open water, the big trees of unbroken woodland, like those now below us, made it the most dangerous terrain for a forced landing. But George never fretted for long, and soon he spotted the thread of a white caliche road, going Gorda's way, and told me to settle down; if need be he could make it down onto that narrow logging track, easy.

"Might lose a couple feet from the end of each wing . . ." he went on, reaching over to poke me in the ribs.

I felt better until the weather closed in. But that happened only far below, where a dense mass of clouds was gradually cutting off the ground. Around '469 the sky remained a brilliant blue, and for hours, between platinum mushrooms that billowed up thousands of feet beyond our aircraft's ceiling George cut happy arcs through sunny cotton canyons. Then I said we ought to start looking for a way down.

Vose shot me a look. "We're up more

than 10,000 feet. Case you don't remember, that climb's taken a long time."

He jerked his thumb at the gray plateau below my window. Deep in its vapor, 70 miles to my right Gorda was stroking across Guatemala. I looked at the murky expanse in which she was hidden, then examined the horizon, every inch of which was carpeted with a blanket of dark wool. It looked innocuous, but that was an illusion.

"That tranquil cloud-world . . . so harmless and simple, became a pitfall," wrote Saint-Exupéry, who, besides being an artist and writer was a pioneer aviator. "Below that sea of clouds lay eternity."

George wasn't concerned. "We got a good signal up here; goin' down now would be foolish. Lose our Neena for sure."

Fleecy cumulus mountains moved slowly past both wingtips, but the farther we flew the more apprehensive I became. Ahead, a narrow rift slit the mantle of wet fog below, and in relief I pointed it out.

Vose shook his head.

"This stuff's getting thicker," I snapped. "What if, farther on, there's not another hole?"

He shrugged. "Go 'til we hit the coast."

That did it. "So what are you planning when we fly right past the coast — which there is no possible damn way we can see through this black ceiling — and end up a hundred miles at sea?"

"Never saw a snowstorm —"

"Vose!" I cut him off. "We have been through that snowstorm crap fifty times, and it has always — always — been baloney. This isn't Maine; these goddamn clouds could go on. . . ."

I waved at their expanse. "Forever. Hundreds of miles."

Vose locked his jaw and said nothing.

By now the rift had become a good-size opening in the cloud bank; through it, far below, was jungle, dark with rain.

"Look: waiting to see — maybe — if we can get down somewhere else. . . ." The idea seemed as crazy as Russian roulette. "That's just crap-ass stupid."

For the first time George's crinkly blue eyes were cold, and they swept over me in a long, hard look. But I wasn't going to let that break in the clouds go by. Face just as frozen, I leaned toward him and pulled rank.

"Long as I am buying the gas, we are going in. Now. Under this cloud bank."

And so, angrier at each other than we

had ever been, through that ragged gap that unfolded into the autumn rains pounding the forests of Central America, Vose and I flew down into Belize.

31.

Lords of Ladyville

Under the clouds, squalls buffeted '469, and with Gorda now lost from the radio, tadpoles of rain squirmed across our windscreen. But I could see the ground, which was what I needed most, and as we dropped over rust-brown Blue Creek and entered English-speaking Belize, Crooked Tree Lake loomed through the mist. We were low enough to notice that the tops of its submerged rain-forest trees were draped with white egrets and black, serpent-necked neotropical cormorants. Through binoculars I tried to get a better look, and what I saw were jabirus. Along the shore, four of the big black-and-white storks stood like sooty-headed pylons, stolidly waiting out the rain. Defining birds of the Tropics, they were a symbol of how far we had come, and I pointed them out to Vose, who muttered that symbols were all we were going to have, now that we'd lost our only peregrine.

Beyond the jabirus, along Crooked

Tree's shore, grew a mat of low woodland completely unlike the tall trunks that thrust up from the lake's surface. They were the short, crooked trees of the region's name, and for 400 years they had been the principal economic asset of this part of the world. Known as logwood and found only in the southeastern Yucatán Peninsula, the trees' gray heartwood was for centuries the world's only source of hematoxylin, a fixing agent that once set the color — from yellow to black — of almost every fabric dyed in Europe.

Cutting logwood was subsidiary employment for sixteenth-century pirates waiting here for Spanish galleons, and later it supported slavery-bound Africans marooned along this coast by shipwrecks. More recently, because the lake has become a bird sanctuary, the surrounding trees have gone untouched, and from above them George and I could see in the distance the country's only commercial airport 8 miles inland from Belize City.

Both our fuel tanks were low, and there was no other air traffic, so I wondered why we weren't coming down on final approach, but since he was still officially not speaking to me, palms up in inquiry, I caught Vose's eye.

His scowl deepened. "This, Mister, is an international field. Tower-controlled. We have to get clearance to go in."

Looking over his glasses, George didn't quite say, "Like in Canada."

By now the airport control tower was almost under our wheels, though with no radio it was clear we weren't going to get anybody's clearance to land. So Vose circled the airfield, wagging our wings. He mumbled about showing everyone we couldn't talk to them, yet the tower's dark-shaded panes gave no hint of occupancy and I doubted that anyone in there had noticed our distress signal.

Then, on the far side of our circuit I saw a dot in the sky, trailing twin threads of exhaust. It was an airliner, coming in on almost the heading we'd just followed. The big jet seemed barely to move, but at 300 m.p.h. it would be on us in seconds and I grabbed George's shoulder. He followed my pointing hand, and almost at the same moment, from the railing surrounding the tower's windows a dim red light blinked frantically on and off.

With a flick of his wrists Vose flipped '469 over into a sharp right bank that, like a diving Swainson's hawk, swerved us out of the airliner's path. Twisting around to look out our back window, I watched the plane — a short-fuselaged 727 belonging to TACA, Honduras's national airline — blow by us, then drop heavily onto Belize's single runway, firing back billows of touchdown tire smoke.

"We'll get in now," said George, floating the Skyhawk downwind for a turnaround. But before we reached the end of the runway another silver dot slipped from the clouds. Also trailing upward-arcing contrails, it grew into an American Airlines DC-8, and as we also swerved out of its line of approach the big plane barreled past on its assigned course.

We still had to land for gas, and on Vose's next pass I looked for the black lantern box propped on the catwalk's rail. It was manned by a compact fellow who wagged its hinged door-front to show us a barely visible green glow, and before anybody could change his mind, George had us on the ground.

Pulled off on a taxiway, Vose and I, still silently angry at each other, waited for the retribution we were sure our violation of international flight regulations was about to entail. Instead, what we saw was a stubby guy in shorts running toward the plane.

Under my wing he pulled up, puffing.

"Ay guay, hombres! Estavamos shitless. Thought you'd get rammed." He pointed back at the tower railing. "We never used that light before!"

"What about your radar?" Vose growled. "Been waggin' our wings at you. . . ."

With that, the young man's face fell. "Someday, perhaps, our Director Alpuche says we should have it. Until then, I watch our sky."

Quietly, Vose asked, "What's your name, Son?" And when Ricardo Cocom had introduced himself, the old Mainer said with gravity that we wouldn't forget his help and asked if, sometime, he'd like to go up and listen for a radio-tagged peregrine falcon.

Cocom thought maybe one of his buddies might, then led us past a fortuitously lettered sign that read

BELIZE — WELCOME ONE AND ALL

into Customs and Immigration. There, a polite Creole lady in a starched blue uniform gave us entry and declaration forms but showed no interest in the Skyhawk, its registration papers, or our luggage. Then, because I had told Cocom that if we could

530

locate our *halcón* we would be staying, he pointed down a muddy road that led to the only village in the vicinity of the airport.

On my map it was called Ladyville, and midway along the road to it Vose and I realized we were passing a military field station. As we made our way between steaming brown puddles, we could see that behind a high wire fence, draped in camouflage netting, were olive-mottled military tents. Right down to its hand-painted sign

BRITISH OCCUPATION FORCES
BELIZE — CENTRAL AMERICA

the camp was a dead ringer for the set of *M*A*S*H*. Except that instead of medevac helicopters, the shark-fin tails of jungle-painted vertical-takeoff Harrier Jump Jets thrust out of the thorny scrub.

Because of George's bad knees and my own, it took a long time to pass the military complex, but between bordering hedges of hibiscus we finally dragged our muddy bags into a three-story New Orleans–style stucco partly shielded from the heat by disheveled metal shutters. It was the Belize International, according to its literature and a weathered sign hidden be-

hind the shrubbery, although — advanta-
geously situated next to the British Army
base — everyone knew it as the Ladyville.

Inside, barely moving ceiling fans
nudged a haze of ancient dust past two
pretty Creole girls, embedded in a mist of
robust perfume, who sat on benches in the
spare lobby. At the news that Vose and I
had checked in for what seemed to be the
unprecedented duration of maybe a few
days, they both broke into smiles, and at
the clerk's nod they showed us the way to
the bar next door.

Out of the lobby it was a few degrees
cooler, and at the sound of clinking ice
Vose had his wallet out before we reached
the doorway. That might have been a mistake, because before he had gotten all the
way across the room he'd accumulated a
cluster of enthusiastic young people,
mostly ladies. None of their heads reached
higher than his chest, and all of them were
chattering at the top of their lungs.

"What?" George shook his silver hair,
bending to listen. "Don't everbody be
talking at once. . . ."

Watching, I climbed onto a seat at the
end of the bar.

After a moment he straightened,
scowled, and pointed out that he wasn't

having a romantic fling with anybody. But among the girls were also kids — young Belizeans who'd apparently never heard of Mexico's feared Secret Halcónes — and soon, one of the littlest, loudest urchins had more or less attached himself to Vose's elbow.

George looked down. "Son," he sighed, "all I'm tryin to do in here is get me a bottle of gin," and before I knew it he'd dug in his pocket and sent the youngster scrambling out the door with an American ten-dollar bill clutched in his fist.

"Vose," I said. "We are already *in* a bar. You think you're ever going to see that kid again?"

Anger finally gone, from above his little congregation of expectant heads George looked at me with disappointment.

"Mister, that young fella could use a tip. Gin or no gin, it's all the same to me."

In the damp air early the next morning, beneath squads of olive-throated parakeets George and I had an easier walk back to the plane. We nodded to the brightly scarfed women with dustpans who dozed on plastic chairs behind the shut-down airline counters, filled out a rough approximation of our flight plan and left it at the empty departure desk, and, heading out to the plane, eased through a side door past the vacant Customs office.

On the flight apron the only activity was going on around three jungle-stained Cessnas being loaded with string-tied crates so heavy the old planes' cracked rubber nose wheels were levered a foot off the pavement. Then George ran '469 down Belize's deserted runway and pulled us into the air.

Before we'd gone a quarter mile, we were in *The African Queen.* Below our wheels the tannin-brown Belize River wound between jungled banks knitted with lianas and philodendron, while ahead of our skimming shadow a little band of orange-and-black aracaris sprayed out of the foliage, skittered like flying fish above the rolling surface of tree crests, then plunged back into the foliage.

I told Vose they were a sort of mini-toucan — probably a family, since the young from previous years stay on with their parents to help raise younger siblings. Larger, lemon-breasted toucans also lived along the river, plucking fruit with lime-green beaks that are half a foot in length, yet so honeycombed with air pockets that each mandible weighs less than a penny.

Protected as Belize's national "bill bird," the keel-billed toucan occupies a safe-guarded riparian habitat that also preserves the home of every howler monkey, crested guan, and hickatee river turtle that lives along the watercourse.

Away from the riverine forest the vegetation thinned into pine barrens, interspersed with grazing land; then on higher ground that caught more seaborne moisture, it again became tall rain forest. At the Guatemalan border, 50 miles west of the airport and nearly 100 east of where we'd last heard Gorda, there was nothing on the receiver, and both Vose and I realized that during the hours since we had abandoned her in the clouds our Niña could have flown all the way down Belize's Caribbean coast. Maybe gone into Honduras, whose government had never responded to my inquiries about radio-tracking aircraft clearance.

So with grim faces we turned south and for miles heard nothing but static. Nor did any signal come over the scanner when we flew a northern loop up to the Mexican border. By mid-afternoon we were back at the airport, and, this time without the signal lamp, we came straight in. There was nothing to do but work on the plane, so, leaving George enmeshed in the wiring of our still largely non-working radio, I went to talk to Luis Alpuche, chief of flight control.

From Ricardo, now our semi-sidekick, I'd heard that Capitán y Piloto — as George and I had instantly become known — had already been fingered as drug agents. That was why Alpuche had called for an interview. A sturdy guy in his middle forties, Alpuche was not the least offended by our novel entry into his country, but our strange uniforms and antennaed aircraft made us a question mark, and it was part of his job to find out who was using his facility, and for what purpose.

After a long series of questions, Alpuche's intelligent face finally loosened, and he rose from the wide mahogany desk that almost filled his cluttered office under the control tower. Motioning to me to follow, he tapped the door to the storage closet he had ransacked for Cocom's signal lantern. "We are glad you did not die in your first attempt to visit us," he told me, the hint of a smile flickering across his broad lips.

We walked on. But at the stairs Alpuche's face hardened, and his cool brown eyes looked directly into mine.

"I understand that, for science, you

claim to be searching after tres halcónes." He waved out the window at '469. "There may be truth in that." I could see him studying, but not quite believing, even the better of my two bedraggled uniforms. "So we will watch each other." He searched for the right words. "While your . . . unusual investigation continues."

It just couldn't continue after dark, he went on, for, like other Latin countries dotted with ephemeral airstrips such as Lerdo's, legitimate aviation in Belize ended at sunset. Vose and I were also to stay out of Belize International's airspace between 11:30 and 3, when all the commercial traffic came through, because we had no transponder — the activating unit every aircraft carries to render it visible to the tracking screens of military and passenger jets.

Otherwise we were welcome to come and go as we pleased — without, Alpuche added thoughtfully, the formality of filing a flight plan for each of our short radio-tracking hops. I was amazed. After the hostile tenor of Mexico, Alpuche's forthright friendliness seemed almost miraculous. But Belize, I was to learn, was a country filled with miracles.

Back at the plane, Mark Vinson hadn't

managed to capture Vose's attention, so he turned his enormous smile on me.

"Rick say you be needin' a mechanic. Mecánico de radio." He placed a proud hand on his chest. "Don' nobody know planes like Vinson."

Darkly glossy as a diving otter, Mark was from southern Belize. Seventeen in a few months and part Garifuna, he said, the descendant of pirates and shipwrecked Africans. That was probably true, though Mark's avionics ability fell into doubt after an hour's consultation with George failed to draw a sound from our speaker.

What Mark really wanted was to go flying, and since he had tried to help us Vose felt obligated and took him up for the day's final Gorda reconnoiter. There we learned that Mark's other interest was wildlife.

"Dat tiger cat," he began, having squeezed forward from the backseat into the minuscule space between George and me, "she mos' beautiful of all de onimals. Still, my favorite be red tiger. Many live at Alabama."

Alabama was Mark's village, and red tiger was the small, russet-coated Central American mountain lion. Like all other wildcats, in Belizean Creole it is called tiger because throughout Latin America *el tigre* is the jaguar, its name having been drawn — like the creatures that are locally called antelope but are really red brocket deer — from missionary books about Old World wildlife.

In the case of *el tigre*, the name is uncannily true, however, for ancestral jaguars were once the only big panthers. Later they evolved into lions, Asian tigers, and leopards — except in the breakaway landmass of South and Central America, where contemporary jaguars retain much of their original primitive form.

That evening there was no sign of Gorda anywhere in northern Belize, and as we dropped back toward the airport, Mark pointed out a tangle of what he called high bush: lofty forest growing between a pair of oxbows near the Belize River, where I'd already heard the braying calls of howler monkeys. It was another of the country's miracles. Next to the village of Bermudian Landing, a dozen or so subsistence farmers had independently decided to pledge 3 square miles of the river's riparian jungle to what — in the Creole bestiary — was called the Community Baboon Sanctuary. Organized with the help of biologist Robert Horwich, that bit of high bush was home to some 600 black howler monkeys (baboons to Mark and everyone else in Belize) that had formerly been shotgunned for food but were now such a source of pride that the airport's cabbies dragged every passenger they could out to hear the howlers' territorial bellows.

As we came in, still radio-less, George had to pull up for another lap around the field. The day's American and TACA flights had long since departed, but a lead-gray, four-engined cargo plane with a U.S. military insignia was gliding onto the runway.

"Paraquat delivery," said Vinson. "Come in twice each week. From New Orleans."

By the time we had landed and were taxiing past, the big plane's freight bay was being unloaded by three coveralled Belizeans whom Vinson said worked for the U.S. Drug Enforcement Agency. They were rolling 50-gallon drums down a ramp toward a pair of aircraft we had not seen before. One was a Cessna spray plane — except for its midnight-blue paint, a replica of Charlie Westfield's agricultural duster. The other was a Hawker Siddeley British military twin.

Vinson pointed at it.

"Dat de spotter," he said. "Go lookin for

marijuana fields. Crop duster follow wid de goods."

It had been a long day, but as we dragged ourselves into Ladyville's blue cinder-block café, George and I found that everything had changed. Instead of potential clients, the hotel's staff now viewed Vose and me as neighbors, and after everyone had said good evening and asked politely if we'd found any drugs, jolly, dark-skinned Belen brought our menus.

There was chicken: fried. Fried shrimp, and chicken-fried steak. Also fried fish. Vose ordered the chicken.

Belen giggled. "Wrong," she said. "Try again."

George was befuddled, so I tried: "Fried snapper."

Belen cracked up. "No fish today!"

"OK, then, shrimp," I said. She nodded happily, wrote it down, and turned to George.

"I don't like shrimp," he muttered. "All I had in Mexico tasted like they battered it in the sand it came out of. So" — he pinned the word under his finger — "steak."

You'd have thought Vose had caught a burlesque pie in the face. Not only Belen but her husband, Umberto the cook, who had come out of the kitchen to watch, collapsed in mirth. Finally, in sympathy Belen draped her hands over Vose's shoulders, and despite his look of pronounced distaste I ordered us two platters of the day's only actual entree: sand-battered shrimp.

Even with our shutters thrown open to the breeze, I slept only half the night. It wasn't the shrimp, it was the growing certainty that George and I had really lost Gorda. So in the dark before dawn, as he grumbled that it was unnecessary since our peregrine was long gone anyway, I dragged Vose out, past the paraquat notices that had mysteriously appeared on the airport bulletin board, proclaiming that every forest clearing in the northern third of the country not located next to a village was presumed to be a marijuana plot and could be subject to chemical dusting.

On the apron where '469 sat parked, puddles covered the runway, and with the smell of the jungle filling our compact cabin, George and I rose into an enormous, crimson sun. Mighty as no northern sun ever is, it was a reddish disc only as it shouldered its way through the horizon's band of haze, for within seconds it blazed into an invisibly bright segment of the sky that, as the Mayan god Kinich Ahau, had drawn the worship of the forest-living people who had raised the first stone temples to its power.

Even in northern Mexico our Niñas had lived in a boreal firmament whose thinner radiance was tolerable to land-bound beings; over the tropical Caribbean George and I climbed into a universe of light like neither of us had ever seen. Away from the vapor rising out of the rain forest, the waves' glistening surface flamed up so intensely that even with shaded lenses my eyes burned, and reflected from the coastal shallows, the sun lit to dazzling aquamarine the underside of the marine clouds that hung over the offshore keys.

But with Gorda gone, all the magic of that Caribbean sunrise was as nothing.

Maybe, I told George, she had gone out to the barrier islands. It was the only place we hadn't looked. Also worried, he began to corkscrew up, and for minutes we climbed, '469's little engine straining against gravity.

Fifteen miles offshore lay Cay Chapel, Cay Cocker, and to the north the big island of Ambergris. Laboriously, we spiraled the remaining few hundred feet that '469 could climb, hoping to find Gorda somewhere along the islands that marked Belize's barrier reef. But we did not, and an hour later we turned back from the sea, just in time for George, helped by the offshore wind, to reach the nearest island landfall.

32.

Ambergris

Our fuel needles showed "E" on one tank and close to it on the other when we shut off our propeller at the little sand-and-tarmac strip on Ambergris. At dawn our sky had been filled with possibility; now, without Gorda, Vose and I were lost.

Drained, we wandered away from '469 and had made it halfway through the village of San Pedro before I realized the streets were empty. Ambergris's largest settlement, Saint Pete is the quintessential Caribbean party town, yet from its row of bars and restaurants not a footstep or a human voice broke the silence. Like the accidental survivors of some sudden apocalypse, George and I tiptoed along the deserted main thoroughfare. Then, in the distance, I heard music. Reggae.

As it grew louder, advancing down the sand-paved avenue came a mass of people, some wielding instruments. As the group was joined by more revelers, we could see that the whole train was led by a huge black spider that leapt from curb to curb, its hairy forelegs groping just above the head of a tiny Creole girl wearing lacy egg-shell wings.

George and I recoiled.

Then the vanguard of the musicians was on us, with one crepe-ribboned giant pounding me on the back and sloshing a paper cup of rum in my face.

"Drink up, Mon! Spider an' De Fly."

Speechless, Vose and I stared at him.

"Day of the Dead. Entiendes? Halloween!"

The parade was only the beginning. Halloween was Ambergris's Mardi Gras, though the island celebrates that, too. Any excuse for a party, and as the town lit up after dark, Vose and I, still stunned by the loss of Gorda, let ourselves be pulled from bar to bar, party to party, by those who had only heard about our bizarre objective and wanted to meet us for themselves. For the first time in weeks we had no Niña. And there were a dozen girls to dance with.

Yet, as had happened among Veracruz's soaring hawks, most of what I thought about was Jennifer — a world-class dancer — and how much I wished that she was also there. Then, during a cardiac respite off the dance floor I realized that from the club's balcony I was looking at the guest house where I'd been told I could find John Fuller.

Fuller was the pilot who had flown — and crashed with — Alan Rabinowitz during his study of Belize's Cockscomb Basin jaguars, and I wanted to ask him about paraquat. On his porch, Fuller said the defoliant hadn't yet come into use during their jaguar work, but he believed that all five of Belize's native cats, as well as most other wildlife, either are killed by paraquat or abandon the sprayed area. What the stuff did to villagers he didn't know.

Afterward, I found El Piloto in the Barrier Reef Bar, running through his Amelia/ Gorda story for a rapt audience, none of whom believed a word he said. Everyone knew Piloto and Capitán were drug agents. But beyond Vose's circle of listeners, a calmer fellow motioned me to an outlying table. It was Barry Bowen, native Belizean, impresario of the local Belikin Beer, and an accomplished private pilot. He'd been talking to Alpuche.

Bowen told me he admired what we were

doing with peregrines and that wildlife conservation was the keystone of Chan Chich, a nature lodge he had set up at Gallon Jug — named for a Mayan ceramic found there — out on the Guatemalan border. Bowen's more immediate concern, however, was that Vose and I not crash our plane while we were in his country. He knew we'd been operating, as he put it, with something of a rough-running engine, and his solution was for us to fly over to a small mainland airstrip where his personal mechanic, Joe Carnegie, would try to get '469 working properly.

During our complimentary breakfast buffet I told George — who was still bleary from the Barrier Reef but not as bad off as me — the plan. He was glad to go, confidently pointing out that, although Gorda had been our biggest, strongest youngster, the Niña most likely to lead us to her winter home, we still had two radio-tagged peregrines coming down through Mexico.

We just couldn't do much searching until we got the plane fixed. So while Vose headed to Carnegie's shop I rented a bicycle and, lugging the receiver in its basket, set off up the island. I'd gotten as far as the Laguna de Cantena when I saw an osprey. Far above the pterodactyl frigates pasted like Halloween cut-outs on the clouds, the osprey revolved in the seclusion of his height. Dark-winged beyond the opal of his breast and shoulders, he silently folded back his crooked wings, and in the strange posture these big fish-hawks use to dive, swung his gray feet forward, next to his beak. Then, his prey evidently having gone down, the osprey flailed out his long primary feathers and returned to his silent circling.

To my right was the lagoon, asleep except for the sigh of the sea; beside my rusty bike wheels, crystalline wavelets rose an inch, then flipped forward onto Ambergris's coral shore. In the heat, a band of black skimmers endured the sun, squinching their cat-slit eyes against its glare. Their orange-and-black beaks aligned like toy-soldier bayonets, every bird in the flock faced the flat water, where another pair of skimmers flew nose to tail, dipping their blade-like lower mandibles to slice long razor cuts in its surface.

Then all the skimmers were in the air, flailing their wings inside a wheeling cauldron of gulls, terns, and skittering puffs of shorebirds. I couldn't tell what had disturbed them. But only one thing could bring such life to that painted scape of sand and sea, and at last I saw the flicker of falcon wings. Pausing to soar on the momentum of their downward sweeps, those almost idle wings were not hunting, and as their owner came across Reef Point, the bowed-out strand where Ambergris protrudes farthest to sea, the peregrine parted a flurry of frightened gulls, then swerved serenely through the halo of empty space their struggling bodies had left behind.

Its carbon-black crown and steel-gray back said it was an adult, therefore not Gorda. Maybe Anukiat. But the receiver was silent, and as she drew nearer I saw she was too large for a tiercel, with broad wings that paddled the air over the bright-blue shallows. For a moment, her broad Kabuki cheeks brought back the arctic steppes, where I'd seen so many such faces, then the falcon was past, and I realized that, except for the scattering gulls, she would have no good hunting on Ambergris: like Padre's dune grass, behind the beach the island's spine was rigged with a long hump of mangroves.

Their thick branches were full of birds, but that tangle was a fortress no peregrine could penetrate, and this falcon was only passing through. So were the small passerines fluttering from branch to branch, but in the mangroves they had found what they needed and would not leave until they'd fed.

Most of them had come in during the night from across the Gulf, riding the southerly air flowing out of a high-pressure zone that had shoved aside the storms of the last few days. Successfully arrived, a male rose-breasted grosbeak — streaked with brown like a thick-billed sparrow in its off-season plumage — sat on top of the thicket, and with admiration for its long crossing, I pressed into the limbs' understory. There the shade reverberated with the broken chirps of a white-eyed vireo, and searching the leaves for it I noticed a basketball-size lump of what looked like brown papier-mâché.

Dappled with shadows, the rough-surfaced sphere fused three forked branches, down whose trunks trailed woolly brown veins — the arched-roof tunnels blind termite workers use to carry their meals of cellulose up to their fortress nest. Above the termites, as tame with hunger as they'd been at High Island, black-and-white, magnolia, and yellow warblers of the mangrove race found only on these cays burrowed into every leafy cluster for the insects that were their food and water. From several feet away I watched their brisk, detailed searches; then below the warblers came a hint of slower movement.

A trickle of viscous sap. But its ooze was too dense for sap, and as I stared, that faintly stirring cord became the back of a green vine snake. Molded as tightly to its russet branch as a parasitic plant, the snake's whip-like body ended in a radically flattened head shaped to replicate the single leaf sprouting at the end of a growing vine. Protruding like every other germinating mangrove shoot, the snake's long nose swayed gently: the tip of an exploratory tendril barely moving in the breeze.

With a quick grab I caught it.

Instantly, what had seemed to be an inanimate stem became a writhing strand of steel. Twisting its body into wiry loops, with a hiss the snake flashed out widespread jaws. But that gaping mouth never closed. *Oxybelis*'s display is mostly show, and as its foreparts stilled into an erect, cobra-like stance a foot from my face, I could see, hidden in the pink mucosa of its open throat, a hint of rear fangs, for vine snakes are mildly venomous. But their mild venom is a threat only to their lizard prey, and since even tiny songbirds were comparatively safe in its presence, I was glad to slide this one back into the branches.

By midday Vose was back, bringing in an uncharacteristically smooth-running '469, and by late afternoon we were again deep in the interior, over rain forest, listening for our peregrines. There, as we climbed through 7,000 feet, I heard an electronic cheep. It might have come from the north, though it was so faint I didn't get a real direction. But the numbers said it was Gorda.

I shouted at George that she must have lost her way in that god-forsaken cloud bank that could have killed us, too. Vose was so ecstatic he let that go, pointing out, with Gorda now again on the wing, vectored east, that he hadn't ever believed Guatemala was her final destination.

From the sectional charts on my lap I saw that if she maintained her speed and heading, before dark she would enter the northwestern quarter of Belize. And for once I was right: at dusk she came sailing out of the rain forest's permanent mist, crossed over the citrus groves that give Orange Walk District its name, and slanted down above the red marsh grass that stretched all the way to the inlet-notched coastal lagunas. From there, on the horizon I could see the offshore islands I had for so long imagined as our tundra falcons' winter home.

But Gorda wasn't going there.

Below, a stand of salt-bleached pines picketed a mudflat where flights of egret and ibis, stilt and dowitcher rose into the air, wheeled, then settled back onto fresh feeding terrain; into those dead pines Gorda's signal descended.

An abandoned osprey nest gave us a landmark, and after circling long enough to be sure she was down for the night, George and I, not daring to rile Alpuche by staying out longer, turned back to Belize International. With the clear telemetry line we now had, Gorda remained on the receiver all the way to the airport and we landed with our lovely, recovered Niña's signal still strong in our headsets.

In celebration, that night George again went for Ladyville's chicken. Suppressing a giggle, Belen shook her head and waited.

But Vose had had enough. "Look here, Ma'am. You got it on the menu. I don't care if somebody's gotta kill the hen, 'cause back in Machias — that's Maine — everybody had chick. . . ."

But Belen was gone. Arms crossed in determination, George sat expectantly. Finally, since I was also hungry and hadn't yet ordered, I went to see about the slaughter. The yard was empty. There were no chickens, and out by the silent road stood Umberto, thumb held high.

Belen padded over. "Car come soon," she said softly. "Only tree miles to de market."

Heading back, Belen hoisted her floppy blue house slippers over a galvanized pipe that ran down the slope of the hen-less side yard, then sprigged into an algae-covered slough.

"Where we gets our water," she said. "No more haulin for de whole hotel."

I was glad I'd stuck to soft drinks and Belikin. Not Vose: at the table I found him knocking back a new cocktail consisting of the tabletop marmalade stirred 50-50 with gin and the opaque tap water pumped from Umberto's new hydraulic plumbing.

"Swamp water's gonna kill you, Piloto; it's straight out of the creek."

As usual, George was sanguine. In front of the light he rotated the fifth of Gilbey's that had accompanied us to the table.

"Drinkin' fellas claim this stuff's an anti-septic. Hundred-plus proof: do in any bug that gets in your system." Gingerly, he touched his queasy stomach. " 'Course, I always kind of doubted those guys."

At first light, Vose and I were back over the pines. Unlike Amelia, Gorda had never been an early riser, and she was still on her roost. George was delighted, and waving at our fuel gauges, right tank still full from our last reconnaissance stop at Chetumal, he said we were as ready to follow her down this Caribbean coast as we'd been to stay with her along the whole eastern shore of Mexico.

But Gorda was through going south. At 8 a.m. her signal winnowed into the air and circled the flats, which after yester-day's long flight should have been a morning search for prey. Then her pulsing cheep turned away from the feeding waders, and from the shore.

It was half an hour before I knew for sure. Miles past Ambergris, climbing with all her might, Gorda was headed straight out over the Caribbean. I didn't panic. With those laser eyes, she had to see that nothing — nothing — lay ahead. George agreed she'd soon realize her mistake and turn back, and for a while our hopes rose. As we reached 9,500 feet and leveled off, minute by minute Gorda's signal grew stronger as '469 began to close on her. But the strain of our full-throttle climb had brought back the Skyhawk's stutter, and even through my earphones I could feel the cabin begin to vibrate. Then, instantly, the shaking stopped. The tachometer flipped back around to zero, leaving our scarred silver propeller windmilling emptily.

Big-eyed, I looked at Vose.

"Water in the fuel," he muttered. "Chetumal. Goddamn Mexican gas."

Swiveling, I searched the horizon. Nothing. Nor was there any sense of falling. Yet we were going down, and fast.

"We'll switch tanks," said George.

He reached down to the floor and twisted the fuel selector valve.

The thing, I recalled, that didn't always work. Meanwhile, the white pointer of our vertical speed indicator swept steadily counterclockwise, counting off the precious feet separating us from the sea.

Then I saw land. Off to my right, a mile down, a strip of mud breached the waves like the back of a surfacing whale. No more than a foot above the water, it looked long enough to land on, and I jerked the yoke over so Vose could see it.

"What the hell you doing?" he snapped, yanking us back toward the mainland.

"*Why not?*" I yelled. "That sandbar's our only chance. Otherwise we're dead."

George peered through the windshield. "It's mud," he said. "Not sand."

Every second we were 80 feet lower and farther from that bit of earth.

"Vose," I screamed. "There's water in our tanks, and we are going down!"

"No," he answered under his breath, and tried the ignition. It cranked twice weakly, then stopped.

I waved at the horizon. "It's twenty miles to the cays; everything else is water."

George shook his head. "I'm not gonna do it."

He really had gone off the end. Maybe I could grab the controls and ditch us close enough to swim to the sandbar. But we were still too high.

The next few minutes were going to determine whether we lived, yet as the wind's roar grew louder, Vose just sat there, silently hunched, twisting the tank selector on the floor.

"Battery's down," he said. "Give it a second. Good gas from the other tank'll run in there."

It was a stupid, pointless way to die, and I was beside myself with rage.

"Mister," Vose offered at last. "I'd lose her. Four six nine'd never get off that bar."

Grabbing his shoulders, I exploded. *"So fucking what? We'd be on land. We'd be alive!"*

George was a big guy, but he didn't resist.

"I wouldn't have a plane," he almost whispered. "Ever again."

I stopped yelling. Without his wings, there'd be no flight-office audiences, no prospects beyond a tin trailer set way the hell out in the desert. Suddenly I saw Vose, slumped on one of his adobe walls, next to an empty airstrip, gazing up.

Cherokee.

I took a breath. This was our deal, then. Too far, now, to make it back to the sandbar even if we turned around, there were no other options. Below my gauzy, scratched-up window, in the indigo depths double-bed-size manta rays circled in pairs. November, I thought. Must be their breeding season.

In minutes we'd be among them, and like gun-shot hawks sucked from the sky, as '469 came down all we could do was face the inevitable.

I reached under my seat and pulled loose a chunk of mealy cushion foam.

"Rotten old seats won't even float."

George shrugged, gave the throttle three or four priming strokes, and once more hit the starter.

Silence.

I could see sargassum weed floating on the surface. At least the sea was calm, and I unlatched my door for a quick get-out.

"Close that thing up," Vose ordered. "Got to float as long as possible."

"We go in," I said bitterly, "I may not be able to pull your stubborn old ass out of here."

"Alan," Vose said, cold as stone. "I don't want you to."

He punched the ignition again, and, amazingly, the engine coughed and spun a single revolution.

"See?" George said grimly. "Mexican water in our right tank. Now hold on; don't want to flood it."

Delicately, he worked the gas in and out, then with one big, trembling hand reached for the starter.

Instantly, the motor was running. Vibration filled the cockpit, and before I realized it we were flying. Our good, left-side fuel tank was feeding the engine, the propeller had dissolved into a silver disc, and the wind was back under our wings — a wonderful, solid viscosity we could rest our craft on and, rising away from the watery void, climb like the side of a hill.

As we went up, suddenly Gorda was back on the radio. Limp with relief, George and I exchanged grins: loud and clear, our familiar youngster, foremost member of our little band of guiding angels, was back with us. But she was pulling higher: a dot too small to see, drilling into the vast thunderheads building across our still-brilliant sky.

Almost simultaneously, Vose and I realized she was gone. Now eighty miles beyond the outermost islands, Gorda had already ceased to be part of our world. Her wild northern spirit, whose imagined dream had drawn us all this way, had at last carried her beyond '469's reach, and high among the Caribbean's towering cumulus, we listened for a long time as her signal waned.

Finally, George looked over.

"Alan," he said, "we'll never know."

Then our brave arctic girl vanished, beyond the radio horizon, into the unknown.

The next morning George and I couldn't

bring ourselves to go up. Into a world of endless sea, Gorda had flown away, leaving a fertile, bird-rich coast where she could hunt, to die somewhere out among the waves. Vose said we had to cut out being silly, but from the charts it was clear that the only land in her path was a speck called Isla Santanilla. Chanced upon by Columbus in 1502, it was more than 200 miles at sea and too small a target for Gorda to hit, even if she knew it was there.

Now she was gone, but as always, confidence continued to flow from George. Finally he pointed out — as though we had any idea if it were true — that since Gorda had followed this route south Delgada and Anukiat couldn't be far behind, and late in the day we inched up into a silent sky.

As the Mayan temples of Caracol appeared below, seeming no more than furry green hillocks overcome by the jungle's verdant erosion, ahead loomed the Maya Mountains, site of Mark's village Alabama. Above it, Vose made two more big loops, both without a signal. We were running out of territory, for beyond the mountains lay the Rio Sarstoon, Belize's southern boundary. This was almost untouched country, where Sharon Matola, the

Belizean Jane Goodall, was attempting to set up a sanctuary for the handful of harpy eagles still found in the tall bush.

I was sure that none of the Guatemalan fighters would find us even if we crossed their country's remote frontier. But there was no point in doing so, because George had managed to coax '469 up to nearly 11,000 feet — which it could manage only in this oxygen-rich sea air — and our listening range now stretched farther down the Caribbean than either Delgada or Anukiat could have flown even if they had come through while we had the plane on the ground.

So we turned back and flew for nearly an hour, hearing nothing but static. Then, just beyond the point where the Rio Bravo flows out of Guatemala's big Chocop wetland, we got a signal. Onto the receiver's little window clicked a pulse so faint I could barely read its #.683.

Delgada!

Holding up the crystal readout for him to see, I yelled at George that our skinny Niña girl had made it — come all the way through Mexico. Incredible as it seemed, that huge-eyed creature who had hissed so savagely at me, even as her gaunt frame had trembled with her capture — that fragile being I believed must have died not far past La Pesca — had been stronger than her fleet-winged prey. Some of it, anyway; alone, she had made enough kills, since we'd last seen her, to have fueled a journey of more than 700 wandering miles across totally unfamiliar terrain.

She was still deep in Guatemala, though, and during the night what Fuller told me about paraquat came back, and I awoke thinking about the drug notices, tacked up despite the fact that few *milperos* — Mayan farmers — can read, have ever been to the international airport, or would have any way to evacuate their little slash-and-burn fields even if they knew they were about to be sprayed with defoliant. As we waited for the three old Cessnas to struggle into the air on their cargo run to Chetumal, Vose remained optimistic, but as we turned toward the spot where we'd last heard Delgada, there was nothing. Maybe she had gone offshore, I said, though I knew she hadn't. All night I'd had the receiver on, with an antenna rigged on top of the hotel. If Delgada had continued on her heading across northern Belize I'd have picked her up.

At a loss, George finally made up his

mind. As we had done in Texas behind Amelia, he announced that we'd start at Gallon Jug, where Delgada's signal had wavered in from the Chocop marsh, then swing larger and larger climbing circles until we either found her or the Guatemalan jets came out.

Twenty miles beyond the Belizean border we were high, but I was on edge, glassing the horizon for military contrails when I spotted a plane. Right down at treetop level, through the binoculars I could see it was not a silver fighter. It was yellow-white, smeared with the stain of jungle airstrips. Beyond it were two identical craft: our familiar cargo Cessnas, on their way — not to Mexican Chetumal, where it was legal to go, but to Flores, Guatemala, where their televisions, stereos, and hard liquor were available only at doubled, government-tariff prices.

I pointed out their little convoy and George banked us around for a better look. Then he gestured through the windscreen. It was only a speck, but the binoculars' lenses brought a knot to my throat. Also just above the trees floated the gray Hawker Siddeley; behind it, drawing a wedge of white vapor across the forest, was another pinpoint set of wings. Wings that through the glasses I could see were midnight blue.

"Paraquat," spat Vose. "That's what happened to our Neena."

As we circled, covering a silent reception area of nearly 20,000 square miles, we watched the spotter plane nose its way from *milpa* patch to tiny forest clearing, leading the blue sprayer that followed, dusting acre after acre of Belizean and Guatemalan jungle. Then the ag plane's deadly plume tapered and, having spent its payload, it turned back toward Belize City.

Defeated, angry, George and I headed back too, though by the time we landed the spray plane had filled its tanks and left on another run. I had read that, as a quaternary nitrogen herbicide, paraquat was more toxic to mammals than to birds, but I also remembered from the California Land Management study that a peregrine found dead in Santa Rosa had concentrated such a high level of toxins from the less severely poisoned birds it had been eating that the government lab told us we should not have even touched its plumage.

The same thing, I was sure, could have happened to Delgada. Feeding on only a couple of paraquat-contaminated birds could have brought her down, crippled or dying somewhere in the forest, her transmitter drowned by the flood the fall monsoon had spread beneath the trees.

Next to the defoliant depot, Vose and I tied down the Skyhawk in our assigned parking space. Heading back to Ladyville, we'd just passed the WELCOME ONE AND ALL sign when a howling siren and the squeal of big tires sent us galloping off the taxiway. It was the airport's old fire truck, laden with volunteers, and it looked like it was headed for '469. But it careened on by and pulled up at the paraquat terminal, where from at least one leaking drum an oily lake had started to spread.

"Gotta get my plane outta there," Vose blurted, but before he'd gone ten steps Alpuche's security men were between us and the Skyhawk. Blown by the wind, the toxin's fumes made it impossible to go closer anyway, so for nearly an hour we watched as the fire squad pumped truckload after truckload of water onto the growing spill. Midway through the commotion, both the gray spotter and the dark-blue ag plane landed, taxiing to a discreet stop upwind of their supply center while the airport firemen — now covered with oily mist — managed only to slosh the defoliant's free-radical toxins over '469, the squatting fuselages of the three old cargo-runners, and finally a considerable portion of the airport's south side parking aprons.

"Delgada's revenge," I said miserably, to no one but myself.

33.

April the Tapir

By the next afternoon it was clear that, text-book toxicity notwithstanding, the airport spill had exceeded paraquat's lethal dosage to birds; the whole southern side of the airfield was strewn with the bodies of great-tailed grackles. Yet with '469 still reeking, George and I went up anyway. As we did the next day and the day after that, over forest still desolate of Delgada. Anukiat was out there too, somewhere, we told each other, and when the spraying crew at last moved on our hopes rose that he might make it through.

Nevertheless, after nearly a week Vose and I had heard nothing but our scanner's empty static, and bitter at seeing swaths of poisoned leaves curl into the gray of mid-winter northern hardwoods, we decided to spend every bit of the next day away from the airport, more or less in protest against the red-letter celebration honoring the arrival of the Commander of the British

Forces — paraquat-spraying partner of the U.S. DEA.

We just didn't stay out long enough.

Returning from our listening zone between Gallon Jug and Blue Creek, I spotted the executive level C-5 dropping out of broken cirrus over the Caribbean. Escorted by a pair of local Harriers, the big military transport let down its flaps and was almost under '469's wings when its tires belched touchdown smoke, which was the signal for its accompanying fighters, saving the best for last, to slam open their throttles.

Blasting out of the slow trajectory they had held over the length of Belize's extended runway, the Harriers' Rolls-Royce turbofans kicked in 23,000 pounds of thrust, rearing their camouflage-painted snouts in the Jump Jets' trademark vertical climb.

In an instant they were on us.

Directly above them, Vose wagged our wings like a madman. But each Harrier pilot was focused solely on the precision trajectory of his partner, and as the fighters came up there was no way to dive away from them. For an instant we could see the dark bulges of the air intakes that gaped like distended gills behind each pilot's clear bubble, and George and I knew we were dead.

Neither of us yelled, but as I shut my eyes for the impact I glimpsed the jet on my side jerk its needle nose off to the right. Then I thought we'd been hit: flung sideways, jerked left and right like a skidding car, battered as hard as in a traffic collision. Yet the cabin was intact, and I could tell that George and I were still up, flying, even though our windows were filled with the smoke streaming back from the Harriers' afterburners.

Once more, I watched Vose shake so hard he took his hands off the controls and simply sat, looking at nothing.

"Twenty feet," he said. "Over here. On your side, maybe thirty."

That was it. Alpuche wasn't going to let '469 off his runway again without a radar transponder — the official reason the fighters failed to see us — though it was really Rick, faithful flight controller Rick Cocom, who had saved our lives. Tracking our "bird" search with his radar eyes, Rick had screamed a radio warning to the Harriers just before impact.

At Carnegie's hangar we found the Skyhawk's flight surfaces were basically intact, though a couple of fuselage panels had been bowed by the jet's exhaust. Joe said he could repair them with rivets, and by the next day George seemed to be OK, too, settled in at the old airfield swapping aviation scuttlebutt with Carnegie, and with Barry Bowen when he flew in from Gallon Jug.

I still hoped Anukiat would come across Yucatán, and I wasn't going to let our receiver languish in a repair shop where it would miss his signal. So I asked Carnegie if there was another plane I could charter to fly out along the country's western border. He said if I was all that anxious to go to Guatemala I should just drive.

Mark Vinson thought he could line up a car, maybe the retired taxi being resurrected by his soccer-star pal Enrique Gonzales; all I'd have to do was buy gas and any parts we might need. But the next morning, with Gonzales's rusty, off-white Olds idling in front of the Ladyville, Mark was nowhere to be seen, so Enrique and I took off without him.

Enrique was a muscular mulatto kid, and his soccer matches were his sole topic of conversation. We had made it through four or five recent replays when, as we motored past a grassy pasture next to the Baboon

Sanctuary, I spotted hovering wings. So slender they bore the glider-like proportions of the nightjars, they were nevertheless wings of speed and power. I shouted for us to stop in time to get the field glasses on the silkily cupped strokes that held an aplomado lightly in place, 40 feet above the ground.

Intent on something hidden in the grass, below her terra-cotta belly the little falcon stretched out her yellow legs and, trimming her wings to a narrower arc, hovered lower. At the sight of her black-masked face helicoptering down on them, the tiny birds she'd been watching bolted, bursting out of the straw in a spray of wings and tails. Blue-black grassquits and collared seedeaters, migrant indigo buntings and yellow-rumped warblers scattered in four or five directions. But the falcon had only 20 feet to drop, and without the speed of a stoop to overtake them, every passerine made it safely into the field's brushy borders.

"Ésta," I said. "Nuestra halcón como ésta." (Something like this is what we're looking for.)

Gonzales was confident. "Entiendo, Capitán. Mi tío sabe bien estes pájaros."

At Hattieville, Enrique's Olds reached

the Western Highway to Guatemala, where we pulled over for a stop at the Belize Zoo. It was another of the country's miracles.

A former film-documentary set, the zoo itself was almost invisible. Within a stand of stubby pines, gumbo-limbos, hog plum, and Billy Webb trees, big unroofed pens enclosed most of the region's fauna, letting the animals live in environments so natural that they hardly knew they were fenced in. Many, like the chachalacas and guans, simply chose to live in the security of the compound, where they were fed by visitors — some of them kids bussed out from Belize City who, despite living in the Tropics, had never seen anything wilder than a pigeon.

Throughout the zoo's first two years its star had been April, a Baird's tapir — named for the same nineteenth-century Smithsonian director as Baird's rat snake, sparrow, and sandpiper. Like a little elephant, she stretched her rubbery snout over the fence to curl my hand in her prehensile lips. A bunch of school kids were headed up the path, each bringing her a wad of lettuce, a carrot, or grass, but April found my gift of greenery first. Then, after a delicate sniff she deftly lipped and tongued each child's offering, taking only a nibble here and there, because being picky was her instinctual style.

Woodland animals adapted to foraging on rain-forest plants, tapirs learned long ago to counter the caustic compounds jungle vegetation evolved as a defense against herbivores. Snipping no more than a few leaves from each of their myriad forage plants, tapirs manage to spread the forest's various floral toxins thinly enough for them to live with only mild, chronic gastric distress, and after the kids had left, April came back to again root through my fingers in hopes of finding a yet wider variety of interesting food choices.

Watching, Enrique had a different take. Poking the plump haunches April pressed against her fence, he pointed out that "mountain cow" was the best eating of all bush animals. But Gonzales was still twentieth-century Belize, for among the park's younger visitors views like his were almost single-handedly being overturned by the zoo's founder and guiding light, Sharon Matola.

As a biologist, Matola knew that tapirs, like the jaguars that preyed on them, were relict creatures, evolved in the jungles of the Miocene and precious far beyond the notion that they were a cheap source of meat. Sharon also knew that to save her country's mountain cows, at least one of them had to have both a name and a media-friendly personality; helped by April's winning ways, in a stroke of conservation genius she had managed to get Baird's tapir designated as Belize's protected National Animal.

Besides tapirs and harpy eagles, Sharon was also interested in falcons, and she told me about a canyon up on Mountain Pine Ridge, not far from the Macal Valley refuge of Belize's few hundred remaining scarlet macaws, where she had seen peregrines; with her sketched-out map, Gonzales thought he could find it.

Not all was Pooh's Corner with Sharon's zoo and sanctuaries, however: I'd read the pro-timbering editorials that attacked her for interfering with the mahogany trade, and the letters claiming she wanted to save macaws only as exhibits for her zoo. As Enrique and I pulled out, near the entrance we saw that someone had hung the body of a big pit viper on the fence.

"Barba amarilla," Gonzales spat. "Yellow-beard tommygoff: kill you terrible." Neither of us thought the snake, a fer-de-lance, had been there when we arrived, and Enrique said it might have been left by someone who didn't approve of Matola displaying venomous *culebras* above signs that explained how snakes were a valuable part of nature.

"In Belize," Sharon had told me, "the one thing you don't want to be reincarnated as is a snake," and despite co-authoring her county's definitive book on serpents she expected little funding would be forthcoming for the zoo's proposed reptile house.

Miles past the Guatemalan line my scanner still hummed emptily, but because the right side of our Olds was riding low, Enrique said we should stop and buy a tire. That was all right with me, even though I'd be paying for it, because next to the wrecking yard was a shop I wanted to visit. Its suspended placard said that the building's first floor was a pharmacy, run by a young Creole man with good English. What had attracted me, though, were the drawings on his sign. One was a mortar and pestle, the other a coiled viper.

Upstairs, the clerk was happy to show me the pharmacy's herbal adjunct, for here, as on the Belize side of the border, jungle medicine was becoming a cottage industry. Sparked by the fascination of European and American media, herbal cures that native society had long relied on had expanded into a burgeoning business. Throughout the backcountry, local guides led medicine walks geared to New Age tourists, showing off curative plants like the thorn trees beside which supposedly invariably grew their antidotal antithesis: softwood trunks whose pulpy fibers could stanch bleeding caused by the thorns of their taller neighbors. This tale was actually more or less true, for the forest was so full of softly fibered plants that one of them was never far from any wound.

A darker side of Central American folk remedies was embodied by the crudely painted viper. Culebra. It was the terror of every rural farmer, chicle-sap gatherer, and cane cutter, and I asked the clerk to introduce me to the snake doctor it represented. He said the man was out, but everybody revered him because he had saved people in nearby villages who otherwise would surely have died.

I knew a little about how he operated. The fer-de-lance hanging from Matola's fence was one of the world's most deadly pit vipers. Yet, like all its kind, it seldom strikes non-attacking animals too large to eat except in a sort of defensive slap whose objective is to drive them away, which means that some of those bitten are not seriously poisoned. Most of the success of snake doctoring, in fact, comes from the practice, followed throughout Central America, of calling every harmless serpent "tommygoff," which means that the majority of the patients seen by the snake doctors have only been nipped by harmless species.

It was the same odds-playing I'd seen among Appalachia's snake-handling faithful: many of their demonstration serpents are non-venomous, and even among the ones that are, few of the careful handlers are actually bitten. Fewer still receive serious poisoning, because their knowledge of the behavioral psychology of serpents lets them avoid truly riled-up animals.

I was disappointed not to meet this practitioner, but as the clerk and I started down the steps, a large, brusque woman heaved up them, pushed past, and announced herself as the doctor's wife and assistant. Because everybody in Belize seemed to have heard I was interested in snakes as well as my clearly fictitious *halcónes*, she tried to sell me, for twenty-five Belizean dollars, a snakebite kit.

"No hospital," she admonished. "From

hospital you gone walk away crippled." That was to some extent true. In the enormous doses necessary to offset serious poisoning, the old horse-antibody antivenin obtained from the United States proved so likely to bring on anaphylactic shock that local hospital medics often chose the evil they knew and simply amputated heavily envenomed limbs.

Wrapped in opaque plastic, the doctor's packet held three different herbs to boil and rub over myself, chewing tobacco, and cigars to either smoke in the woods or boil into a repellent poultice. Pit vipers have a keen sense of smell, and for some reason everyone in Belize believed they hated the scent of tobacco. The only things of value in the kit were two loosely woven scarves to wrap around a bite, for such mildly compressive bands would squeeze the lymph system — viper venom's primary dispersal route — just enough to temporarily impede the toxins' circulation.

Down on the street Gonzales was waiting, with the bill, beside three new tires. Around him fidgeted a crowd of uneasy young men as varied in shape and color as pebbles in a stony creek. All of them, including the two who were

Enrique's cousins, seemed to know exactly why we were here, because while Vose's and my three weeks of semi-mysterious reconnaissance flights were old news by now, it was assumed that my trip to Guatemala signaled some fresh investigation, perhaps one that would generate a market for valuable information. So, moving away from the others, Gonzales's tallest, shiftiest cousin motioned me over.

"Maybe," he hissed, "I be knowin' de place to look for your bird." He glanced around to see if anyone was watching, which of course everyone was. "Up by Orange Walk Town," he went on, palm just barely open. "Could be sunrise. Tomorrow."

Away from town, through cutover forest, Enrique and I drove along drying clay roads blocked here and there with huge transport trucks, still mired to the axles from last week's downpour. At the village of Enrique's relatives our falcon contact turned out to be his uncle Julio, a parrot dealer, whose house and all its many porches were filled with bedraggled red-lored, white-crowned, and yellow-headed Amazons. Some of the adults had been caught in nets, but most were the unsold remnants of last breeding season's young — fledglings retrieved from cavities in the chainsawed trunks of hollow trees.

Since those babies were fed almost exclusively on the deficient diet of human-chewed corn to tame them, no more than a third of those who survived their first plummet onto the jungle floor subsequently lived to be shipped out as pets, and none would ever return to the wild. Yet their loss was minimal compared with the destruction of the giant rain-forest trees in whose hollows Amazons lay their eggs.

For Matola's cherished macaws, the problem was worse. Not only was a nestling worth more than a man could make in half a year of agricultural labor, but the only trees massive enough to contain the huge nest holes macaws require are the giant mahoganies, matapalos, and kapoks — trees now so few in number that even deep in the forest they were easily located by trappers like Tio Julio.

Julio was as disgruntled that I was not the customer he'd expected as I was at seeing his parrot mart, and because 20 miles past Melchor de Mencos the road to Tikal became impassable, it seemed unlikely, with my receiver still buzzing emptily, that Enrique and I would get a reading from Anukiat unless he flew directly above the forest canopy. To maximize our antenna range we needed to be high, I decided, so I stuck Sharon's Pine Ridge map in front of Gonzales and told him to turn around.

34.

Xunantunich

Hours later, as we lurched up hillsides of rutted red clay, I glassed the sky for the loose banner tails of the scattered scarlet macaws that still ranged along the upper Macal Valley. If any were visible, I wanted to spot them first and divert Gonzales's attention so he wouldn't have anything to report to his uncle. But only a handful of turkey vultures circled, brushing their backs against the underside of re-forming rain clouds. No black vultures were among them; those not feeding among the garbage dumps of San Ignacio were watching from down in the trees, waiting to fall in behind the larger scavengers if the turkey vultures began dropping toward a hidden carcass they'd located by scent.

From the canyon's rim, I looked in vain for the pale body-stockinged king vultures that also sometimes flew with the turkeys, then, with the receiver on SCAN, sat down to wait. In ten minutes Enrique was bored;

in twenty he had stacked his new tires beside the Olds and was stretched out fast asleep on the backseat.

It seemed strange to be searching for a barren-ground arctic peregrine above wet tropical forest whose nearly 400 lofty tree species could not have been more different from the bonsai world of every *tundrius*'s birthplace, and I had no idea if the falcons Sharon had seen here were even peregrines. But one of our far northern Niñas had come almost this far and I'd seen another tundra peregrine out at Reef Point. Yet, as the wind gusted up over the valley's steep edge, with a panorama of cloudy sky to pull down signals from, those details vanished in the spectacle below, and once more I was the demi-falcon I'd been in Alaska, transfixed by the endless flight of birds passing through the gorge below.

As a swarm of Vaux's swifts, erratic windup toys, cut abrupt trajectories above the gorge at my back, the pines murmured in the rising breeze, then breathed more sharply, as if a mammal had scurried along a branch. Turning, I saw there was no squirrel: the rustle had been alighting wings, and on the highest limb a perfect little falcon bent forward, scrutinizing me with yellow-ringed eyes that shone like smoky glass. From his black head and egg-shell throat I thought for a second he might be a tiercel peregrine; then I noticed the rust of his breast and legs and his exceptionally long-toed yellow feet and knew he was the far rarer orange-breasted falcon of the Tropics.

He bobbed his head, and with a staccato patter fell backward, as if blown off his perch, disappearing behind the pines. Thirty seconds later he burst out of the forest below, scattering the swifts and keeping up his angry chatter as he swept down the glen toward its outlet above the Sibun River Gorge.

In fifteen minutes he was back, directly under my overlook, sailing the opposite way up the valley on wings the color of granite, mottled with darker crescents. But it was going to rain, and as the first sheets of mist slid down the mountain, buoyant as a kite the little falcon flared up toward the far side of the canyon, bowed his wings to break his speed, and vanished into the shadows of its vine-laced cliffs.

By the time Gonzales and I reached the turnoff to Xunantunich, the rain had let up and broken moonlight marbled every clearing along the road. Gonzales wasn't happy about stopping anywhere after dark, but I wanted to see the ruins under the tropical moon they had, in part, been built to worship. So while he waited in the car, leery of the tommygoffs he said had bitten two people here, I paddled a tied-up skiff across the Belize River and walked the wooded path to the old temple.

On its great white steps, swept by flakes of slithering cloud shadow, I tried to see the people who had set those stones. Theirs had been an intense culture, risen to power on the strength of its shade-sheltered agriculture of maize, beans, and cacao, all grown on gravelly limestone that is among the least-fertile substrates ever farmed by a major civilization. But this humid isthmus had reserves of fresh water — springing from the ground, rushing down its rivers, and falling from heavy seasonal rain — that for centuries allowed the Mopan Maya to clear and then irrigate ever-larger swaths of jungle. Finally, their enclosed fields' produce let an expanding population quarry the huge limestone blocks they used to raise towering sacred cities where, before, men had built only with sticks and thatch.

Xunantunich's main temple had long ago been cleared of vegetation and archaeologically inventoried, so, scrambling from one level to the next, it was easy to make my way up. On the ceremonial platform at its summit the silver moon was brighter, the surrounding forest a dim smudge of vegetation silently aflutter with the membranous wings of bats. From here, Xunantunich's rulers had waged war, using every ounce of will and cunning to escape being enslaved by their neighboring city-states, staging royal ball games fatal for the losing team, and engaging in the ritualistic penile bloodletting of kings who were also obliged to function as symbols of fertility.

Yet, for all the power of their society the rulers of Xunantunich eventually faltered. The flow of water they had manacled so well, leading it here and there with elaborately engineered supply channels and stone-walled growing paddies, eventually broke its chains. Like other cultures, that of the Mayans could operate only within the framework of its own experience, which held no hint of the decades of drought yet to come. As dwindling crops exacerbated the civil wars already raging, famine felled the cities, leaving the last residents of the Mayans' vast farming communities to return to their earlier jungle lives of hunting and growing *milpa* crops, abandoning their high stone temples to the swift colonization of strangler figs and once-domesticated cohune palms.

Silent now of sacrificial screams, the old plazas swelled with gentler voices. Rising and falling, from the surrounding trees came crescendos of trills, clucks, eek-eek-ekks, and guttural wonks. It was a chorus the earth knew long before birdsong, or even before the hiss and roar of the great reptiles. It was the nocturnal symphony of frogs.

Hyla, agalychnis. Tiny beings, slender-legged because they seldom had to leap from danger, they were treefrogs: mostly pistachio green, with a pale patch on their sides called a "flash stripe" since it's supposed to confuse predators, more than 100 species live in Central America. Each has its own distinctive call, most of which seemed to be simultaneously pouring from the forest around Xunantunich.

Other ears were more discerning. Like leathery moths, predatory bats swished past my face, among them fringe-lipped frog eaters that cut unerringly through the enveloping constellation of amphibian sound. With the tips of their delicate ear-shells cocked to guide them to the individual croaks of their prey, they plunged with sonar-guided confidence into the inky slots between the jungle branches, and I thought that, on this stormy November night, the old sacrificial platform of Xunantunich was the most wonderful place I could be.

But with my receiver still humming emptily in search of Anukiat's long-vanished signal, I had to wonder why I was really here.

Since leaving Padre I had seen a lot of birds, most of them — except for the jabirus at Crooked Tree and the orange-breasted falcon that had found me perched on its rimrock — comparatively ordinary ones. But mine had never been a birder's quest, a search for exceptional species. Instead, Vose and I had managed to share the lives of a few falcons, had learned exactly where at least one sub-adult tundra peregrine had gone after she flew south from Texas's barrier islands, and had discovered much of how she lived along the way.

Yet neither of us was an ornithologist, a lifetime member of the Peregrine Club, or even a falconer, and the environmental havoc George and I had seen around Campeche's oil tanks was no more than what anyone with an eye for nature would have noticed just by passing through.

If not that, then what?

I thought awhile, noting that as the clouds thickened, closing off the moon, some of the hylas fell silent while other species moved stridently into their emptied vocal space. Finally, I decided that Vose and I, each in our separate way, had sought the same thing that had drawn me, on that first airplane ride with Janis, to peregrines.

It was not the depth of the umber eyes glowing from our Niñas' cheeks, gold as the rising moon. Nor was it the predatory dives that obsessed Riddle and his falconers. It wasn't peregrines' incandescent fierceness, or even Amelia's continent-spanning geographic wisdom and endurance on the wing. It was nothing that could be seen.

"Anything essential is invisible to the eyes," I remembered from Saint-Exupéry, for from the first what had crystallized Amelia in my heart was her invisible drive, her all-encompassing quest for home. It was the same search, for some impossibly remote winter asylum, that had subsequently drawn Vose and me to our now-lost Niñas.

"Trackin' peregrines — best thing I've

ever done," I remembered George blurting as Gorda, confirming she was alive and again on the wing, had risen from the Papaloapan swamps. Like me, he seemed to have found some psychic mooring here, on the road, in the air, going up every day with the never-ceasing goal of falling in behind a swiftly moving falcon.

Why else would Vose have come this far, and with such determination? Only because of something so essential it was almost indiscernible. Aloft, George was still a bird himself: a flier, youthful as he'd ever been, yet wise with the experience of his years.

The real dream, I saw again, had been ours. The vision that by joining our peregrines' ancient journey we could somehow become part of what Edward Abbey called the heroism and grandeur of life, the hidden struggle of the million avian lives that, all around us, were enduring what it would seem could not be endured. Pushing on through storm and famine, downpour and sudden predation — always with the homeward-streaming determination that, for some of them, would at last overcome the barrier of inconceivable distance.

It was a testimony to the optimism of life

itself, and it was what Delgada, even lost in the paraquat forest, and Gorda, now flown away to sea, had given George and me. It was the grail I had sought long ago, on the prairie marshes of my boyhood — a time when I'd had no real home and had longed to join the cranes, join the geese, join every hawk and skittering shorebird whose breast burned with the power of return to the distant harbors I also sought.

But reaching back to that awareness had been costly. Since those youthful dreams, other events had come into my life, and into the life of another, caring person. Jen.

I thought of Henry James, *The Beast in the Jungle*, and its character John Marcher, who guarded too long his imagined visitation from the mythic jungle cat he both revered and feared; waited too long to commit himself to the seemingly lesser experience of the love of May Bartram — which "in that cold April twilight . . . was perhaps even then recoverable."

I had done as much with Jennifer. Had left her, willing to board our little plane and face whatever might befall it. I had shut off the same chance for a life both emotionally riskier and far less grand than the one I had tried to find on the wing, with Vose, spinning our way up and down the continents, betrothed to the myth of our mission's biologic significance.

Between black rags of cloud, moon shadows swam across the temple clearing, and in them I saw that Marcher's fatal choice had become my own black beast. Bats swooped, and the chorus of treefrogs went on as before. But everything was different. My heart racing with the realization of what I'd done, somewhere in the darkness a real jungle cat coughed its guttural snarl. I knew it was a jaguar and hoped it would come closer, but back by the river Enrique had heard it, too, and through a quarter-mile of treetops I saw his headlights come on, killing the night and drawing me down the temple's dark path, back to Ladyville.

The Cockscomb Range and Mountain Pine Ridge were the highest points in Belize, but the tallest place in the northern part of the country, where Gorda had come out of the Yucatán, was the airport control tower. There, while '469 waited for new parts to arrive, with one of the plane's big antennas clamped to the building's rooftop radio tower, I spent my days, thanks to Alpuche's generosity, listening for Anukiat behind the control room's shaded glass.

From there, as Cocom, still dreaming of his promised radar, scanned the sky like a nest-bound falcon, I gazed down at the human traffic that flowed across the runway. Every morning the three liquor-laden Cessnas still filed their bogus flight plan for Chetumal, took off, and a mile beyond our sight turned left for the Guatemalan border. Paternally, Rick followed their course with his field glasses and, never mentioning a word to his boss, waited expectantly for them to return each afternoon.

At noon and 12:30 the TACA and American flights came in, emptied their bellies of travelers headed for the offshore beaches, and took on a load of their homeward-bound counterparts. Twice a day I climbed the radio beacon pole to check the antenna's stainless-steel prongs, certain that by now Anukiat must have either crossed the Mexican isthmus to the Pacific or, perhaps riding the orographic flows like Amelia, had worked his way down the mountainous spine of Central America.

But after Xunantunich, Anukiat didn't matter so much. Things were different now, and every day I tried to reach Jennifer. Apologize, tell her our life together was still recoverable. Ask her to join

Vose and me. But all I got were empty rings or her recording, and each afternoon I was tempted, if it hadn't meant abandoning George, to board that departing American flight to Houston.

I was on the stairs, heading back to the hotel, when Rick came running.

"Alan. Come quickly: we have you in our flight pattern!"

And there, circling the field, wings wagging in pride flew Vose, proving that once and for all we were reignitioned — ready to land and take off in the company of every DC-8 and 747 in the skies since, as George said after he set '469 down, Carnegie had tuned, rewired, and mechanically massaged the Skyhawk enough to get rid of its stutter, and that we had better pack because he'd come up with a plan.

Anukiat hadn't gone anywhere, George thought, at least not probably. He was just taking his time coming down through Mexico. And even if he had somehow gone by — well, we'd just missed him and that was that. But the core of his idea, George said, was that we would take the plane as high as possible, retracing our route back through Mexico until we found him.

What about the Yucatán, I said. Those 300 miles of roadless jungle?

Gotta go that way anyhow, George waved. Plus there was always that logging track we could get down on if we had to. Since Xunantunich, that was the kind of reasoning I didn't even want to resist, and the next morning Vose and I were off the runway ahead of the smugglers.

As we turned northwest, the Caribbean swung away behind my window. I hadn't expected to feel anything, but as rain-forest clouds blurred the shoreline's vanishing peacock shoals, a wave of sadness swept over me. Cloaked in my headphones, concentrating on picking up even a faint transmitter cheep, suddenly I realized that I'd never again fly with Vose over those lazuli waves, never know the unselfconscious acceptance of Cocom and Alpuche, egocentric Enrique, Belen and Umberto. Even the friendship of Mark, *el gran mecánico,* now returned from his mysterious absence.

What awaited us in Texas I did not know. But Vose and I had come a long way, and I felt sure that here we were close to finding our own lost wintering ground. To turn back now, before the seasonal tide of life flowed north again, seemed a failure, a foregone chance to grasp a dream that had been almost within our grasp.

35.

Beyond the Sea

It was an hour before the engine began to miss. Just a little, but Carnegie's tune-up had only temporarily resolved the Cessna's chronic stutter, and across Yucatán's endless sweep of trees Vose and I sputtered north toward darkening clouds — for this was the season when the first winter storms deepen the turbulence of the autumn monsoon. To get above the impending weather George started to climb, stutter and all. But even at an altitude that would let us hear Anukiat anywhere from Chetumal to southern Guatemala there was nothing on the receiver, and after refueling at Villahermosa, before dark we once more reached the Sierra de Tuxtla. Both tanks were low, though before we went in I asked Vose to make a quick pass over the mountains' ridgeline.

And there he was. My little guy. For weeks I had imagined Anukiat continuing to hunt near where we'd seen him last, not far south of the Rio Grande, but with his signal cheeping as cheerily as if we had never lost him, our little arctic tiercel had reached the Tropics after all. On the slender wings I'd rocketed into the air behind Morizot's cabin, I felt he had come to meet us. It was an emotion bereft of reason, but still I couldn't speak. Just showed his numbers to Vose — who didn't catch on for a minute. Then, all the way around a joyous aerial circle, George beamed proudly, for we both knew that, once again, the old flier's instincts had brought us to our falcon.

"Probably been here all along," he announced, stabbing a finger down at Cerro Santa Marta, the Tuxtlas's highest prominence, where Anukiat had gone to roost.

Still circling, with our headsets off, Vose and I tried to figure what had happened. Somewhere up by La Pesca, I thought, Anukiat must have waited out the tropical storm, then come through Veracruz's funnel of raptors with later waves of soaring hawks. Only long after Gorda had he arrived in the Tuxtlas, and now we would learn if the horn of Mexico was really his wintering ground.

"That Gorda — she stayed here two whole days," Vose recalled. "I bet this one's home." But if he was not, with our destinies once more linked I knew that nothing would stop Vose from locking in behind the charcoal-blue blades of Anukiat's wings: back across Yucatán, over Belize's tropical forest, and if he chose that course, as far out to sea as we could fly.

Then, with our guy down for the night, Vose set the Skyhawk in on Minatitlán's familiar landing field, where we knew there was a good cantina and a nearby place to stay.

It was a huge mistake.

The airfield's gas pumps, both its runways, and the surrounding grassy fields were uncharacteristically empty. While Vose checked '469 over I poked around and found a frightened teenager sequestered in a concrete-block storeroom. Reluctantly, he agreed to unlock the hose and pump our gas, then from atop '469's wing he bent to confide in me.

"Muchos se han matados," he hissed. (Many have been killed.) As I listened in growing horror, I gathered that the day before a shoot-out had occurred between the village's home guards and an armed band of what the local guys had taken for smugglers trying to ship a boatload of marijuana down the Coatzacoalcos River to a freighter waiting in the Gulf. Faced with the superior weaponry of real soldiers, more than a dozen of the town's defenders had simply been annihilated. Everyone still alive was in shock.

Finished with our gas, the boy climbed down, glared at me, and, money in his pocket, waved at the far end of Minatitlán's long runway.

"Vete! Muy pronto," he ordered. (Get the hell out.)

But we could not. Storm clouds sweeping in from the Tuxtlas had hidden the mountains we had to bridge, and to wait out the rain Vose and I retreated to our familiar cantina.

Except that now it was no refuge; in half an hour the door banged open, and three quasi-commandos, none with matching combat gear but all carrying automatic rifles wrapped in black foam rubber, stalked in. Everyone froze. Never raising the muzzles of their guns, with stony eyes the trio glanced around the room, then fixed their gaze on George and me.

Humble as captives, we bent silently over our plates. There, the incongruous thought came to me that had we been religious men Vose and I would have been praying, and that if we had been, I knew what mine would have entailed. They'd have been thanks that after weeks in the Tropics George and I had finally stowed our soggy, *halcón*-emblemed uniforms and were back in jeans and T-shirts.

Evidently not seeing us as a threat, the little militia's leader grunted, and as one his followers turned and barged back out. For a long time nobody moved, or spoke, and finally Vose and I just left money on the table.

By then it was too dark to take off, even without the rain, so we shoved the Sky-hawk's seats as far back on their tracks as they'd go and spent a sleepless night planning to lift off, no matter what the weather, if anyone approached. As far as we could see — which was only a few yards — nobody did, and soon after sunrise I heard the concussive pulse of heavy helicopters.

To our surprise, the federal soldiers that poured out of the olive-green troop carriers paid no attention to us: their immediate objective was the flotilla of journalists just starting to arrive from Mexico City. Every writer was trying to get into Minatitlán to talk to the wailing women we'd seen on the road, but, like corralled sheep, each approaching carload of newspeople was immediately penned into a roped-off enclosure set up on the airport taxiway.

Most of the journalists spoke English, and from them George and I learned that their assignment was to cover what had already been deemed the Minatitlán Massacre. It was no picnic. As the fog burned away, the tropical sun beat down, and the writers, exposed in their open enclosure, were miserable. There was no water, the women were desperate for *baños,* and rumors flew: besides the slain local militia — no one knew exactly how many — the helicopter troops had supposedly shot four bandits. Then we heard that they weren't really bandits but Guatemalan revolutionaries, and finally someone said that the morning's gunfire hadn't hit anyone at all.

With the sun, however, George and I saw our chance. Carefully, we eased away from the still-impounded journalists and headed for '469, parked as far down the runway as we had been able move it the night before. This time there was no pre-flight; in two minutes Vose and I were rolling across the concrete, passing military choppers that we dared not look toward for fear of catching the hostile eye of one of their guards.

★ ★ ★

Aloft, I clicked on the receiver. Anukiat was gone, and a big listening loop showed he was nowhere for a hundred miles in any direction. Flown on across the Yucatán like Gorda, I theorized, though I was beginning to realize how silly it was, at this point, to still be concerned with a wandering falcon. Yet, because tracking peregrines was what George and I did, listening to our receiver all the way, we flew back to Belize.

From the airport, I reached Don Morizot. One of our group's Padre-banded falcons — an adult female, but not Amelia — had been trapped at Recife, out on the eastern tip of Brazil, 4,000 miles from where Gorda had left Central America for the sea. It seemed barely possible, but maybe our strong young Niña had known where she was going. Perhaps she had just kept on, all the way across the Caribbean. One December, in guiding birders through the Antilles, I'd seen a tundra peregrine looping around a rocky point, and with only short gaps of ocean separating those islands, their volcanic peaks formed a chain of landfalls that ran all the way to South America.

Maybe Gorda had managed to reach those close-set emerald summits. Anukiat, too. Just as he had done along the entire eastern coast of Mexico, our strong arctic tiercel might be following the same invisible path. That supposition was supported by the fact that, for the next five days, even with scans the length and breadth of Belize, as well as far out over the offshore keys, Vose and I were unable to pick up his signal. George was sure Anukiat had simply reached the Caribbean ahead of us and that — along with Gorda and Morizot's adult female — he would eventually reach Brazil. In which case, because Vose had never been particularly upset over any of the Skyhawk's problems, he said he'd be more than willing to press on after them.

He even — as usual — had a plan. A new magneto would arrive from Miami any day, and Carnegie could rewire one of our old ones as a backup. Then, to cross the thousand miles of open ocean that lay between us and the nearest South American shore, George and I were to wedge a 50-gallon drum of aviation fuel into the Cessna's backseat and run hoses, through holes cut in its Plexiglas rear windows, up to the fuel inlets on top of the wings. With those makeshift lines taped in place, and me working a beer-keg hand pump, the two of us could theoretically resupply our gas tanks with enough high-test to make it.

Finally, I thought, I understood the entirety of what it meant to be George Vose — and to face the fact that, although he was probably willing to ultimately go down in '469, I was not.

What I told George was the truth. Even if the Skyhawk should by some miracle make it to South America, there was no way the two of us could. I was seriously fatigued, and after taking us off Vose spent long periods doubled over with intestinal cramps while I steered the plane.

Then — dialing collect, since we were virtually out of money — I made the rest of our calls. Vose's mother was unchanged, but my wonderful grandmother had died. And a pocketful of change brought me only a recorded message from Jen, telling me not to call.

I kept trying, though, tying up the airport's single pay phone, and finally I got her at work. There was relief in Jen's voice when she heard me, but I knew I had to talk fast.

"We lost our peregrine; the last one. Plane's waiting for parts, and when they get here you've got to come down and fly home with us. Vose wants you to have more lessons."

Before I could get any further she cut in.

"Sweetie, it's November — which, knowing you and George, neither of you may be aware of."

She gave a little laugh that I took as a positive sign.

"I am *so* glad you're all right. And that George is."

She paused, and I could feel the warmth vanishing.

"Not that I had any way of knowing. All I knew was that your lawyer — Binder — when I called, he said he'd try to find you. Every American in Belize is supposed to hang out at the Fort George Hotel, but they'd never heard of you."

"Rich tourist joint. But listen: I'm done. We've got to drink some coladas; there's great places to eat, right on the water. We can really talk."

It was too late.

"Hush, now," said Jen. "If you keep on I'll have to hang up. Without going over everything."

Trying to keep her from going over everything, I said I knew how angry she was, but Jen told me she'd gotten over that weeks ago, after I had tried to phone from somewhere — Villahermosa — then left only a ten-second message.

A planeload of tourists was pushing down the corridor, and left hand over my open ear, I nodded as they squeezed by.

I said it was all right, that what I wanted now was just to settle down. No more risks. Which, after what I'd realized at Xunantunich, was true. But Jen didn't believe me.

"What it is, Hon . . . is that you *won't* take risks. Emotional risks."

I looked at the ceiling: another of her weird damn takes on life. But this one was about to sink me. I punched the wall, drawing the attention of Bill Wilkes, Alpuche's big security guard. But I didn't care.

"Sweetie," Jen went on, and 700 miles away I could see her shaking her long-suffering head. "It's only what you told me. Told me so passionately I'll never forget it. About that enormous long trip your baby falcons had to make. Leave their nest bluffs for . . ."

"That perilous journey."

"Their migration. You said it was to discover what was inside them."

Whatever that meant, at least we were talking; all I wanted was to keep her on the line.

"Sure, I know," I agreed. "It's the most affecting thing I've ever seen."

I hesitated. "But Vose and I are through with that. I'm back with you."

"No, you are not. You are still way out there. So far out I'm not sure you even know who you are anymore."

She waited, expecting me to understand.

"Hon," she said at last, "who would I be taking back?"

Jen was crying, and I saw Marcher's black beast rise, huge and hideous, for the leap that was to settle it all. "That's why both of us have to go our separate, perilous ways."

Dazed, I hung on.

"Those big risks," Jen said. "They're the only way you can find your way." She paused. "Your way back."

There was a short silence, then she hung up.

I walked away from the dangling phone. Everything in the terminal seemed strange. Wilkes started to come over, but I told him I just needed coffee, aiming the feet I felt hardly connected to out the side door that bypassed Customs, into the parking lot where Vose, still sick as a dog, was waiting.

Three days later we reached the Rio Grande. There, almost the last thing

George and I saw before we crossed the border was Guzman's silver Beechcraft, crash-landed in a Mexican coastal marsh, its aluminum fuselage ripped by gunshots. Then, with no place else to go, late in the afternoon Vose set '469 down at Cameron. Not 50 feet from where Janis had introduced us, George, wobbly with parasites, shuffled his feet.

"Ought to get back; truck some water to my little cottonwoods."

I knew he would get well, would go on, and that one day both of us would run faster, stretch out our arms farther, until . . .

Until we realized our dream was already behind us. Back along the Bay of Campeche or out among the coral cays. Deep in Belize's emerald jungle, or on the road to Ladyville.

Yet most of that dream — as inaccessible, now, as long-lost Amelia — lay beyond the sea. There, Gorda could already be flicking her deadly wings above some equatorial delta, a marshy shore where the golden plover would just have landed. A place where the continent's long sweep toward Africa ends and the Amazon pumps its silty blood a hundred miles into the South Atlantic. It was a sanctuary that still seemed impossible for our falcons to reach, but now I too would have to go on, alone, in search of my own distant harbor.

As I tossed my duffel into the truck I heard our old motor cough, miss, and come to life. Then '469 was in the air, heading west into the still-bright evening sky. Soon, I knew, Vose would trim its angle of ascent and fold his big wrinkled hands over his lap, and feeling lost that I was not on board, for a long time I watched the little plane rise. Then, so far off I could barely make him out, below the first small stars the old skyhawk dropped a notch and cut two tight circles, wagging his wings good-bye.

Epilogue

Ten miles past the beachfront condos, Padre's flats look the same as ever. Spring and fall, peregrines still sweep across their sandy expanses, and I still wait for them there.

More than a decade after Ken Riddle and his colleagues discovered the tidal flats' tundra falcon population, another large, mostly unknown group of peregrines — many of them also arctic *tundrius* — has turned up along the Gulf Coast, recalling for Vose and me that one of the northern adolescents we initially radio-tracked turned back to the Texas beaches after less than a day's flight into Mexico.

We assumed she was one of those youngsters who simply never make it to their southern wintering grounds. Yet I've observed both adult *tundrius* and *anatum* falcons wintering along the Texas and northern Mexican coasts. Moreover, it is now believed that a significant group of *tundrius* spends the winter offshore.

In spite of the overall negative impact of the oil rigs that sit on the Gulf's shallow continental shelf as far as 130 miles from land, their platforms have become important rest stops for ocean-transiting passerines. For falcons, these offshore buildings are perfect hunting sites: during October, and then again in spring, up to four northern peregrines wait in the steel superstructure of many of the offshore rigs.

That is a large number of falcons — a group that might constitute a sizable percentage of the arctic-peregrine population Riddle initially discovered. Opportunistic as ever, the oil-rig falcons (which also include many merlins, as well as one carefully documented aplomado) sit on the platforms' elevated perches, chupping angrily at fishing boats that intrude into their predatory territory, then sweep out like flycatchers to intercept the small passerines crossing miles of open water. With no cover except the oil rigs' superstructures, small migrants sometimes manage to hide among the machinery, but sooner or later every individual must leave the steel girders for the sea.

At that point they are easy prey for the falcons, who from strategic sites on the platforms' superstructures often launch themselves — frequently flying, wingstroke for wingstroke like the mated pairs I saw along the Colville — in twosomes. No one knows if these birds are mates, temporary associates, or even generally of opposite genders, but they are clearly together, calling back and forth, chasing each other just off the waves and returning to perches only a few feet apart. Perhaps belying the assumption that after arctic peregrines leave their northern nesting territories they spend the rest of the year in solitary migration, many of the adult oil-rig *tundrius* hunt, chatter together, and may ultimately reach the Tropics as mated pairs.

Some, however, occupy the offshore platforms all winter. Deep in January and February I've seen them out there, lost in what had seemed the heart of winter — until from my fishing boat I recalled the world they had flown away from. A universe of almost galactic hostility, cloaked, at that moment, in what had already been many weeks of unremitting blackness, gale-force winds, and -60-degree cold described in e-mails by Jim Helmericks, dug into his thick-walled home at the Colville's mouth. But out on the winter Gulf, where sunny days often prevail, every passerine and wading shorebird had, nevertheless, long since departed for the Tropics. Still unrecorded is what these oil-rig stalwarts prey on during the frigid months of little migration, but it may be the laughing and smaller half-pound Bonaparte's gulls that venture far from land; the occasional wandering Forster's tern; or even the scoters and goldeneye sea ducks that are light enough for a female peregrine to carry back to eat on a petroleum platform.

Ashore, on Tobin Armstrong's oak savannahs and those of the neighboring King Ranch, aplomado falcons are also making a comeback. Now (due largely to the success of the Peregrine Fund), aplomados are the only endangered falcon remaining in the United States, and since 1993 the fund has been releasing them along the Gulf Coast. More than thirty pairs now nest south of Matagorda Island. A similar program, augmented by a "safe harbor" agreement that exempts release-site ranches from some of the restrictions of the Endangered Species Act, has seen young aplomados hacked — meaning set free under careful supervision — into their old desert prairie habitat on the Miller and Means ranches in West Texas's Marfa Basin, as well as on the McKnight spread east of Marathon.

Moving in a different direction, Ken and Rebecca Riddle left the University of Texas M. D. Anderson Cancer Center and, with

John Hoolihan, sought the hawking realms of Ken's youth by accepting an offer to practice falconry on a Great Khan–like scale, setting up falcon-breeding and raptorial medical facilities for the Emir of Dubai. As I said good-bye to Ken, who was astonished by the amount of funding available for his project and was deep in making long lists of the medical equipment he would need, he was also contemplating packing his prized mammoth tooth from the Colville.

Three thousand miles to the north, that river looks the same, though by the turn of the twenty-first century, I couldn't miss, when flying in, the new drilling rigs and gas-production sites that have sprigged up across the coastal plain. When Riddle's expedition went down the Colville, seismic crews were just beginning to map the Alpine Oil Field, downstream from Nuiqsut, with vehicle trails. These pathways were set on 4-mile centers; now, the next stage of seismic exploration calls for the trails to be narrowed to axes of no more than a half-mile, turning parts of the arctic coast into something that may ultimately resemble huge housing development road grids.

Despite the petrochemical companies'

emphasis on how few acres are physically occupied by their drilling and production platforms — each of which is now much smaller than a few years ago — the National Academy of Sciences Research Council has noted that both the initial seismic exploration and the subsequent extensive support systems necessary to sustain the North Slope's enormous oil and gas industry call for an area of more than 1,000 square miles to be connected by a network of roads, pipelines, drilling pads, gravel pits, and huge manufacturing and living facilities like the one I passed through at Prudhoe Bay — which, along with its eighteen subsidiary oil and natural-gas production areas, has now become North America's largest petrochemical field.

The cumulative result is that in the very heart of the continent's last great wilderness now lies a vast industrial/energy complex. The only terrain along the northern coastal plain not subject to occupation by this commercial development is the Arctic National Wildlife Refuge, an area roughly two thirds the size of the state of New York, with a predominantly Native American population so few in number that most of its members could be seated in a couple of Airbuses parked on a runway at Kennedy Airport. The protection of ANWR has, of course, become a major battleground for conservation groups and the energy companies seeking access to the three to nine billion barrels of oil distributed in pockets beneath the refuge's tundra.

On January 22, 2004, however, the conservationists lost. Interior Secretary Gale Norton approved a plan to open most of a previously protected 8.8-million-acre segment of the Arctic Slope to oil and gas development.

Even with careful management, the presence of so much human activity is likely to be destructive. During the years that I led people to watch the seasonal movements of large land mammals, again and again we saw huge grazing herds brought to a halt by human blockage of their traditional routes. Unable to go anywhere, the few remaining black rhinos no longer migrate; the horse-like kulans of the Gobi Desert face almost certain extinction; and as my clients and I watched, thousands of African wildebeest were forced to abort a 200-mile migration due to the sudden erection of wire-mesh cattle fences.

Among North American grazers, the longest migration is the 1,700-mile round-trip march made by some of the Porcupine herd caribou. The herd travel from the eastern border of Alaska to ANWR's coastal plain every spring to give birth where predators are few and the wind off the Beaufort Sea sweeps away many of the biting insects that plague the young. A few months later the herd, accompanied by its newly mobile calves, normally threads its way back through the foothills of the Brooks Range. Soon, either the host of brightly lit new petroleum platforms or their vast network of support roads and supply depots may permanently interfere with this natural movement.

Birds sometimes have it a bit better. In 2001, eight of my beloved cranes — the endangered, locally resident whoopers — were relocated to Wisconsin's Necedah National Refuge from Florida's small, non-migratory flock. The cranes' introduction to this northern biome, flush with food in summer and free from the hurricanes that threaten the cranes along the Gulf Coast, however, meant the newbies had to learn to travel, which they were able to accomplish only with the help of Operation Migration, a Canadian group that uses an ultralight aircraft. The pilot, dressed in white like a surrogate parent crane, slowly guides the newly flighted chicks — who have listened to the recorded sound of his engine since they lay encased in their eggs — 50 or so miles each day until they reach their winter home in northwestern Florida.

As with just-fledged goslings following their parents down from the Arctic, the juvenile cranes need to be shown the way only once. After that first journey they have evidently internalized the route so thoroughly that, over the last three years, thirty-six chicks have made the round trip entirely on their own.

Upriver from the new oil fields, petroleum-exploration crews have also been working out of Umiat. Motorless, my old aluminum river boat sits abandoned in a patch of weeds, and a recent fire revamped some of the camp's unique architecture, but the place would have been different anyway. Its guiding spirit, O. J. Smith, died of cancer in 1999. Herds of caribou continue to ford the river there, peregrines chatter above the airstrip, and once in a while, beneath the headlands where rough-legged hawks, ravens, and gyrfalcons still nest, a grizzly leaves its tracks along the rocky shore.

Yet out on the Chukchi Sea, the grizzlies' relative, the polar bear, as the climax predator in a long marine food chain, has recently been impacted by polychlorinated biphenyls. These toxins have reached levels of 10 parts per million in the Canadian Arctic, and skyrocketed up to 100 ppm along the northern rim of Europe. There, carried by air and sea currents from both Old World cities and the new northern refineries, PCBs, which act as endocrine disrupters, have been implicated in the simultaneous development, among Norwegian bears, of both male and female sexual organs, prompting media wags to term them bi-polar bears.

And it's not just bears. The breast milk of Inupiat mothers from northern Quebec who eat a great deal of seal meat, often contains PCB concentrations two to ten times higher than that of more southerly women with a less-concentrated marine-mammal diet.

Until recently, such observations were mostly random data points, but during 2003, U.S. medical biomonitoring formally established that almost every American citizen is steadily accumulating, year by year, an ever greater "body burden" of PCBs, DDT, dioxin (now linked to both prostate and skin cancers), and other organochlorines banned decades ago — along with similarly escalating levels of flame retardants (three to ten times higher in the breast tissue of California women than that of Japanese; twenty times higher than in women and infants from Sweden and Norway, where these chemicals are prohibited), uranium, and mercury. South of the Rio Grande, the report went on, DDT derivatives such as DDE are now three times more prevalent in the tissues of Latin Americans than among most North Americans. This confirms my own observation that DDT canisters are part of the agricultural equipment of nearly every rural village between Tamaulipas and Venezuela, leading me to wonder how our friends the Martinez boys are faring.

Among the Arctic Slope peregrines — one of the original environmental barometers of methyl sulfone metabolites from PCBs, DDT — similar poisoning may still be going on. Tundra peregrines were officially removed from federally endangered status in 1994, yet U.S. Fish and Wildlife raptor biologist Ted Swem estimates that only slightly more than half the peregrine pairs currently nesting along the Colville manage to raise young every year, and that the number able to do so may be declining. With eggshell thinning still occurring in the area, this seems to be a paramount reason to establish the National Wilderness Preserve along the Colville that has long been proposed by Tom Cade and Clayton M. White.

Farther south on the Pacific seaboard, peregrines *are* making a comeback in both California and the Columbia River Basin between Washington and Oregon. But this revival may be due largely to population augmentation, in which captive-raised young are being filtered into wild peregrine populations. As in many parts of their range, among peregrines nesting in the western part of the Columbia Basin, eggshell thinning also still prevails; there, more than 1,000 captive-hatched young have been released since 1981, compared with just over 200 offspring fledged from about fifty untouched nests.

Finally, the theory that young arctic peregrines separate permanently from their parents somewhere along their first migration is now further supported, according to Alaska Fish and Game biologist Peter Bente, by ongoing observations that by the

time post-fledgling peregrines reach observation stations in the Canadian and
northern U.S. Rockies they are apparently
fending for themselves.

These days, though, nobody travels with
them. As Vose and I knew would soon be
the case, falcons are now tracked by satellite. In 1993, two peregrines captured at
Assateague Island off the Maryland and
Virginia coasts were fitted with powerful
transmitters whose signals are received by
orbiting satellites. The following year one
of these birds was tracked by the new telemetry to wintering grounds in central
Argentina. Nevertheless, seventeen such
platform transmitter terminals, or PTTs,
attached to peregrines on Assateague,
along with a number of other transmitters
fitted to birds in Greenland, have shown
that some migrating tundra falcons — like
the similar group that winters along the
Gulf Coast of Texas and northern Mexico
— travel no farther south than the Mid-
Atlantic barrier islands.
    Four years later, in Mexico, two of Rid-
dle's former team members, Tom
Maechtle and Bill Seegar, along with Mi-
chael McGrady, Juan Vargas, and M.
Catalina Porras Peña, fitted eleven female

peregrines and one adult male with similar PTT transmitters. These long-range telemetry units let the researchers determine that most of their migrants left the Tamaulipan Coast during the first week in May and took about a month to travel between 2,846 and 3,631 miles, reaching widely separated summering grounds spread between far-western Canada and the coast of Greenland.

These birds' satellite-relay transmitters also allowed them to be monitored during their autumn migration. Beginning in August and September, the falcons followed southward routes through the middle of the continent that, after an average of about forty days, brought them to northern Mexico during September and October — from which wintering ground one of the study's females was ultimately satellite-tracked back to the same summering area the following year.

At about the same time, other groups began using similar orbiting satellite telemetry. In both Edmonton and Wood Buffalo National Park, Alberta, Geoff Holroyd, of the Canadian Wildlife Service, fitted adult *anatum* peregrines with comparable transmitters. Held in place by Teflon webbing crossed over the falcon's chest, this type of transmitter is designed to send signals six hours a day every three days during fall and spring, and to relay each bird's identification number and geographic locale through an Argos ground station for approximately a year.

That kind of chest-webbing-attached backpack seemed to me like a detrimental, airspeed-cutting accouterment, although Holroyd found that, even with the packs, peregrines breeding in Alberta sometimes made the trip to wintering grounds on the Pacific coast of Mexico in as little as eight days and returned the following spring in as few as six. Some traveled down the Caribbean's Lesser Antilles, a route along which I have also observed migrating tundra peregrines. (Despite their strong flight capacity, in 1998 one of Holroyd's falcons was satellite-tracked into Hurricane Mitch and lost in the storm.)

Finally, continuing through the summer of 2002, from the High Arctic Institute at Thule, Greenland, the Peregrine Fund mounted a major study on the ecology and migration patterns of that island's two falcon species, one of which includes the world's most northerly nesting group of peregrines. Since the Peregrine Fund began research on it during the 1970s,

Greenland's peregrine population has grown significantly, with breeding pairs now numbering in the thousands, though the region's gyrfalcons may have decreased.

The latest telemetry tracking of female peregrines from this area involved lighterweight, solar-powered satellite transmitters, which, according to project leader Bill Burnham, leave the falcons free to fly effortlessly. "So far," the Fund's Kurt Burnham wrote in 2003, "it appears that all the birds are wintering in South America."

Yet, even with the global movements of migrating peregrines becoming better known every year, satellites don't tell you what is happening on the ground: show you what a perplexed first-year migrant or a seasoned adult peregrine sees every day, what the weather is, and what the wind sliding across its body feels like to someone traveling in its company. No set of relayed data points can capture the rip and slash of the long wings that Vose and I sometimes saw below — and that for weeks forged on, responding to every change in cloud, and sun, and the chance for prey — a thousand miles from where I'd set them back into the air.

★ ★ ★

After Vose and I came home from Belize, Luis Alpuche, with our left-behind scanner still abuzz on his desktop, could hardly help but have his curiosity piqued by the potential passage of the falcons that we had told him might come within range of his receiver's control-tower antenna.

"I got to where I'd take that radio gear with me every time I went hunting," he told me. "Up on Mountain Pine Ridge they thought I was crazy, or about to fly away in some little plane. But when I came in, I always put your antenna back on the roof and waited a bit longer."

Weeks later, in the vicinity of the airport Alpuche began picking up transmitter signals. Anukiat, maybe even long-vanished Delgada, since her frequency was also still in the receiver's search band. I caught the first flight for Belize. But before I arrived the signals had disappeared, and on numerous subsequent trips, though I found other tundra peregrines, I never picked up one of our falcons' electronic cheeps.

Just as it was when George and I flew in by signal lantern, Belize International flight control still operates without radar. That deficiency eventually became more than Ricardo Cocom could bear, and after writing to George about the sophisticated electronic systems he dreamed of learning to operate, Rick made his way to the United States — where he was apparently so taken with airport technology that, expired visa notwithstanding, Vose told me, he never went home.

Alpuche moved on to a job with American Airlines, and out at the Belize Zoo, Sharon Matola remains the country's foremost conservationist. In addition to the incursions of loggers, however, who for years have wangled timbering leases that whittle away at the high bush Sharon still struggles to preserve, in October 2001 Hurricane Iris hit the Stann Creek District with 140-m.p.h. winds, leveling much of the tropical woodland that remained. In less than two hours Iris destroyed 800,000 acres of coastal broadleaf jungle and upland hardwoods, including a good deal of Alan Rabinowitz's Cockscomb Basin Jaguar Preserve — now formally recognized by the Belizean government and supported by the World Wildlife Fund — as well as, farther south, a swath of harpy eagle habitat. Afterward, reclusive animals like tamanduas, tiny silky anteaters, and tree-less spider monkeys roamed the desolate countryside; emerging from fallen forest along the

Southern Highway I saw a single, dazed black jaguarundi.

But even Iris was a natural disaster that tropical woodland is adapted to overcome, and within a year the jungle had begun to recover. Far more dangerous is Chalillo Dam, currently planned to be built on the Macal River by the Canadian energy company Fortis, Inc. If completed, according to Matola and the Belizean non-governmental conservation group Bacongo, its impoundment will destroy one of the wildest areas left in Central America, home to Belize's last 200 scarlet macaws, to jaguars, tapirs like April, threatened Morelet's crocodiles, and hundreds of avian species, both the northern migrants on which peregrines depend for prey and local forest dwellers.

Matola, Bacongo, and environmentalists including Robert Kennedy, Jr., of the Natural Resources Defense Council and Elizabeth May of the Sierra Club have been joined by celebrities such as Harrison Ford and Cameron Diaz in opposing Fortis's planned construction. Besides Chalillo's devastating effect on wildlife, the loss of the unexcavated Mayan temples it would inundate, and the risk it would pose to nearby villagers, Fortis's monopoly on electrical distribution in Belize gives local rate payers — who are already charged three times what their Mexican neighbors pay — no choice but to subsidize the $30 million to $45 million project. With a minimal capacity of but 5 to 7 megawatts, the dam will produce the most expensive electricity in Central America.

This is the dark side of what Vose and I saw as Belize's magic tropical kingdom. Last year, heavily vested local politicos passed a bill shielding Fortis's ambitions, despite the opposition of the company's own wildlife consultant, the Natural History Museum of London, whose ranking scientist on the project, Dr. Alastair Rogers, wrote Belize's chief environmental officer, Ismael Fabro, to say that "constructing a dam at Chalillo would cause major, irreversible, negative environmental impacts of national and international significance." He added that "no effective mitigation measures would be possible."

The fate of the Macal Valley's forest now lies with the pending decision of the Privy Council — five Downing Street law lords who make up the highest court having jurisdiction over Belize's British Commonwealth. On her return from the court's initial hearing early in 2004, Sharon told me that the Privy Court judges knew they had been lied to — for one thing, the dam's supporting bedrock turned out to be porous limestone instead of solid granite as the court had been told — but she felt their decision ultimately rested on what the judges elected to do with that information, as well as with the Belizean government's determination to proceed with construction despite every ecological warning they had been issued.

With an anti-Fortis editorial by Robert F. Kennedy, Jr., appearing in the Toronto *Star*, both Canadian and world opinion against the project was building. Yet in February 2004, according to the San Padre (Belize) *Sun* it appeared likely that the dam would be built:

> The three-judge majority of the Privy Council acknowledged that the proposed dam would flood an area scientists say is "one of the most biologically rich and diverse regions remaining in Central America." . . . Nevertheless, Lords Hoffman, Rodger, and Sir Leggatt deferred to the Belizean government's political decision to allow the Canadian-backed project to go forward.

★ ★ ★

Finally, George Vose didn't go down with his plane. Cessna '469 was crashed, however, far beyond repair by a student pilot who had borrowed it for his first cross-country flight and miraculously escaped with minor injuries. The insurance was just enough to let George buy 5323 Tango: an even older, less-powerful Skyhawk whose left brake was also barely functioning the last time we went up.

Long before that, however, Vose had returned to Mexico. Working for U.S. Fish and Wildlife, during the mid-1990s he went back to monitor flocks of white-fronted geese that, radio-tagged in the Arctic, migrate to a winter range around Chihuahua's 7,000-foot-high Laguna Bavicora.

"Big gangs of 'em; sandhills, too," he told me during a trip home. "I have to decide whether to fly over them, under them, or zigzag right through the middle."

Vose was seventy when he started the goose project, and Fish and Wildlife thought he'd never stick it out at Bavicora for five whole winters. But he did, despite being snowed in for more than a week on a couple of occasions. Then George hooked up with a bear. A big, black male who wandered into Alpine, Texas, one summer night.

Befuddled by truck headlights and police sirens, the bear climbed a tree next to the county courthouse, where he was shot full of tranquilizers and collared with a transmitter by a crew from Texas Parks and Wildlife. Hauled 90 miles to the Rio Grande, the Courthouse Bear, as the papers called him, was released on the assumption that he'd cross the river into Mexico.

But over the arid borderlands, using Amelia's old receiver, retrieved at last from Belize, Vose (who was over eighty by then) and I flew unsuccessfully in search of him. George maintained that the bear was too old and wise to remain for long in the desolate terrain below, but since we got no radio pulse, he truly seemed to have vanished.

Vose's determination didn't wane, though, and weeks later, on a flight out to his ranch — still ear-muffed to our old scanner — he picked up a signal. Just as he had thought, the bear was working his way north. Headed for the wooded mountain country where he'd been darted, under George's periodic aerial supervision he finally reached the crest of Mount Ord, a

6,800-foot peak that overlooks the county seat. There Vose and I found him again, on pinion- and oak-covered slopes splashed with red autumn leaves.

Probably looking for somewhere to den, we figured, because in the Trans-Pecos bears don't really hibernate, they just lie down in a sheltered nook to sleep for a couple of mid-winter months. Vose stayed too high for us to spot the bear, and on the way back — out of the blue, as always — he leaned over and confided that he hadn't really wanted a look. After so many years doing aerial telemetry, George said he'd come to hate accidentally getting close enough to see the animals he tracked. I told him how a cinnamon-colored bear had fled in panic below a Bureau of Land Management helicopter in which I'd flown to monitor the aeries of California peregrines, and he agreed that nobody had any business causing wildlife that kind of stress.

And thus, with thousands of feet of silent air separating his plane from Mount Ord's limestone canyons, somewhere in northern Brewster County, linked only by radio, George and the Courthouse Bear remain to this day.

Acknowledgments

In the interest of conciseness and in the hope of not trying the reader's patience, I have selected from and in places compressed events and observations that took place in the course of my searches for peregrines from the Far North to the Tropics; the names and personal histories of some individuals have been changed to protect their privacy.

I would like to express my gratitude to those who generously shared with me their knowledge of birds of prey and gave me photographic prints and transparencies of these wonderful creatures. For such photos, information, and help I am indebted to the members of Ken Riddle's and the M. D. Anderson, University of Texas System Cancer Center's Padre Island Peregrine Research team — especially John Hoolihan, Bill Satterfield, Jodie L. Pacy, Bob Whitney, and Janis Chase — as well as to the members of Ken Riddle's Colville River expedition: Rebecca Riddle, Tom J. Cade of the Peregrine Fund, and organizer Skip Ambrose of Alaska Fish and Game.

I am also grateful for the support of Don Morizot of the M. D. Anderson, University of Texas System Cancer Center and his Padre Island researchers. And I am indebted — for the same sort of data and photographic contributions — to the Peregrine Fund, to the World Center for Birds of Prey, to Jim Ince, and to Brian Walton and the members of the Santa Cruz Predatory Bird Research Group.

Also, my great thanks to Austin attorneys Bob Binder and Burrell D. Johnston; to the late O. J. Smith of Umiat, Alaska; to F. Chavez-Ramirez; to Jake Eberts, Christine Whitaker, Adam Leipzig, and Laura Lodin at National Geographic Feature Films; and to Belizeans Luis Alpuche, Ricardo Cocom, Barry Bowen, and Sharon Matola.

Due to the limits of space, not everyone I traveled with is included here, but to everyone who made possible my searches and shared firsthand my fascination with peregrines, I extend my deepest gratitude.

I am also indebted to Ed Acuña, Peter Bente, and Geoff Carroll of Alaska Fish and Game, Kurt Burnham of the Peregrine Fund, Klemie Bryte, Carol Edwards and John Gee, Victor Emanuel, Shawna Fisher, Clare Innes, Erik and Vennie Jendresen,

Karen Kimble, Monte Kirven of the California Bureau of Land Management, Dan Klepper, Greg Lasley, Tom Lehr, Bill Mading, Terrence and Ecky Malick, copy editor Pat Fogarty, production editor Maria Massey, Jonathan McIntyre, the Platte River Whooping Crane Trust, Humphrey V. Ogg, the late George Plimpton, Edward and Annie Pressman, Robert Redford, Gerry Salmon, Irv Schwartz, Larry Smitherman, Ted Swem of U.S. Fish and Wildlife, Karen Tenkhoff and Bill Holderman at Wildwood Enterprises, Daniel E. Thompson, Will Wallace, and Troy Wismar.

More than to anyone else, however, this book owes its existence to the faith and foresight of agent and initial editor David McCormick, of the Collins McCormick Literary Agency, and to the empathetic and perceptive literary judgment of editors Deborah Garrison and Sonny Mehta at Alfred A. Knopf.

About the Author

Alan Tennant was born in Fort Worth, Texas, and grew up in Texas, Florida, and the Caribbean. He has taught film and literary criticism at the University of Texas and has lectured on ecology in more than twenty countries. He is the author of several books on wildlife and nature, including *The Guadalupe Mountains of Texas*, which won the Southern Book Award, Western Books Award, and a best nonfiction award from the Texas Institute of Letters. He currently lives in West Texas and conducts natural history seminars and trips around the world.